MATRIX ALGEBRA
An Introduction

KRISHNAN NAMBOODIRI

Series: Quantitative Applications in the Social Sciences

a SAGE **UNIVERSITY** PAPER

38

Series: Quantitative Applications in the Social Sciences

Series / Number 07-038

MATRIX ALGEBRA

An Introduction

KRISHNAN NAMBOODIRI
University of North Carolina

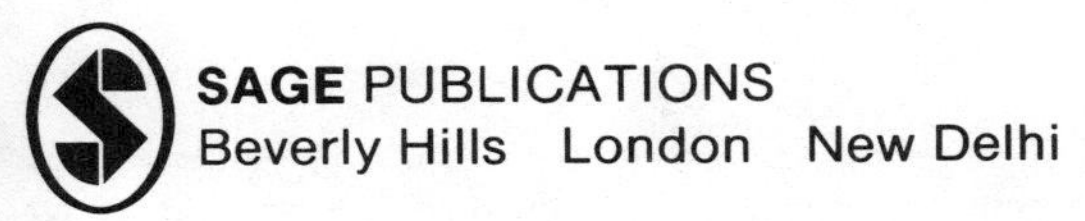

1984

For information address:

SAGE Publications, Inc.
275 South Beverly Drive
Beverly Hills, California 90212

SAGE Publications India Pvt. Ltd.
C-236 Defence Colony
New Delhi 110 024, India

SAGE Publications Ltd
28 Banner Street
London EC1Y 8QE, England

International Standard Book Number 0-8039-2052-0

Library of Congress Catalog Card No. 83-051623

FIRST PRINTING

When citing a professional paper, please use the proper form. Remember to cite the correct Sage University Paper series title and include the paper number. One of the following formats can be adapted (depending on the style manual used):

(1) IVERSEN, GUDMUND R. and NORPOTH, HELMUT (1976) "Analysis of Variance." Sage University Paper series on Quantitative Application in the Social Sciences, 07-001. Beverly Hills and London: Sage Pubns.

OR

(2) Iversen, Gudmund R. and Norpoth, Helmut. 1976. *Analysis of Variance.* Sage University Paper series on Quantitative Applications in the Social Sciences, series no. 07-001. Beverly Hills and London: Sage Pubns.

CONTENTS

Series Editor's Introduction

Matrix algebra is an important tool in mathematical social science and is, in addition, the vocabulary used in teaching elementary and advanced statistics. More and more it is anticipated that social science practitioners have at least a rudimentary understanding of it. Nonetheless, social science majors and graduate students often fail to go far enough in mathematics to get a thorough grounding in the topic. For this reason, we are delighted to publish *Matrix Algebra* by Krishnan Namboodiri.

The reader will find Namboodiri's volume eminently readable. It begins with the most basic definitions, which serve as an illuminating yet relatively complete introduction, even for those who have no previous acquaintance with the topic. Chapter 2 explains elementary manipulations of matrices such as matrix multiplication and inversion. Chapter 3 introduces the concept of linear dependence of rows or columns of a matrix, a concept that is useful when discussing systems of linear equations. Chapter 4 is concerned with the slightly more difficult concepts of eigenvalues and eigenvectors. In each case the ideas are amply illustrated with fully worked out examples.

For anyone wishing to achieve a thorough understanding of both mathematical and statistical analysis of the social sciences, matrix algebra is an essential ingredient. Its smooth presentation makes *Matrix Algebra* by Namboodiri the place to start.

–*Richard G. Niemi*
Series Co-Editor

MATRIX ALGEBRA
An Introduction

KRISHNAN NAMBOODIRI
University of North Carolina

1. INTRODUCTION

Rectangular Arrays

We come across rectangular arrangements of numbers frequently in our everyday readings. Table 1.1 contains an example. In this case, the numbers are the results of the 1982 U.S. Senior Open Golf Championship. The numbers in a row represent the scores of a particular player, and those in a column represent the scores in a particular round. Thus, the position (row and column) of a score in the arrangement identifies the player to whom and the round to which the score applies.

Another example is the weather report we usually find in daily newspapers (see Table 1.2). In this case, each row represents a city, and the columns represent the highest and lowest temperature for a particular day.

We encounter such arrangements in our research work, e.g., responses gathered in an interview survey arranged in the form shown below:

	Interview Question				
Respondent	1	2	3	4	. . . m
1					
2					
3					
4					
.					
.					
.					
n					

Examples from elementary mathematics include detached coefficients and constant terms from a system of equations such as

$$2x + 3x = -1$$
$$x - 2y = 3$$

arranged in the following form:

Equation	Coefficient of x	Coefficient of y	Constant Term
1	2	3	−1
2	1	−2	3

And so on.

We now define a "matrix" as a rectangular array of numbers. Our intention is to treat such arrays as single objects. To explicitly indicate this intention, we enclose the array within brackets as shown in 1.1 below:

$$\begin{bmatrix} 73 & 72 & 74 \\ 62 & 65 & 25 \end{bmatrix} \quad [1.1]$$

Instead of brackets, one may use parentheses or double ruling on both sides as shown in (1.2)

$$\underset{\text{(i)}}{\begin{pmatrix} 1 & 2 & 3 \\ 4 & 5 & 6 \\ 7 & 8 & 9 \end{pmatrix}} \qquad \underset{\text{(ii)}}{\begin{Vmatrix} 10 & 11 & 12 \\ 13 & 14 & 15 \end{Vmatrix}} \quad [1.2]$$

The numbers that constitute a matrix are called the *elements entries* of the matrix. We refer to the elements by their row and column numbers, in that order. Thus, the (2,1) element of the matrix in 1.2 (i) is 4; the (2,3) element of the matrix in 1.2 (ii) is 15; and so on. Obviously, if a matrix has n rows and m columns, it has altogether nm elements.

A matrix that has n rows and m columns is said to be of order n by m or $n \times m$. Thus, the matrix in 1.2 (i) is of order 3 by 3 or 3×3; the one in 1.2(ii) is of order 2 by 3 or 2×3. In giving the order of a matrix, we *always* mention the number of rows first, followed by the number of columns.

TABLE 1.1
Results of the 1982 U.S. Senior Open Golf Championship

Player	*Rounds*			
	1	*2*	*3*	*4*
Miller Barber	72	74	71	65
Gene Littler	73	69	76	68
Don Stokes	75	69	72	70
Bob Goalby	72	71	74	72
Gay Brewer	73	70	75	73
Arnold Palmer	73	71	73	74

We now know that a matrix is a rectangular array of numbers; we have introduced the concept order of a matrix and the convention that when we give the order of a matrix we always mention the number of rows first. If we stop here, we could not claim that matrices are interesting or useful. What makes them useful and interesting is that we can "make them work" for us, under a specific set of rules of operations.

In developing the rules for matrix operations, it is convenient to use a single letter as a label for a matrix and also to use letters to designate its elements so that we may refer to matrices with arbitrary elements. To distinguish between the letter designations of matrices and those of their elements, we shall follow the convention of using boldface capital letters for matrices, and lower-case, ordinary letters for their elements. An example is

$$\mathbf{A} = \begin{bmatrix} a & b \\ c & d \end{bmatrix}$$

Here we use the label **A** (a single letter in boldface) for the 2×2 matrix whose elements are a, b, c, and d. (The equality sign is used here to indicate the label.)

Other ways of using letter designations for the elements of a matrix are shown in 1.3 (i), (ii), and (iii).

$$\underset{\text{(i)}}{\begin{bmatrix} a_1 & a_2 & a_3 \\ b_1 & b_2 & b_3 \\ c_1 & c_2 & c_3 \end{bmatrix}} \quad \underset{\text{(ii)}}{\begin{bmatrix} a_1 & b_1 & c_1 \\ a_2 & b_2 & c_2 \\ a_3 & b_3 & c_3 \end{bmatrix}} \quad \underset{\text{(iii)}}{\begin{bmatrix} a_{11} & a_{12} & a_{13} \\ a_{21} & a_{22} & a_{23} \\ a_{31} & a_{32} & a_{33} \end{bmatrix}} \qquad [1.3]$$

TABLE 1.2
Weather Report for Selected Cities on a Particular Day

	Temperature	
City	*High*	*Low*
Amsterdam	79	68
Athens	82	64
Bangkok	79	77
Barbados	91	79
Beirut	77	68
Belgrade	74	55
Vancouver	75	55

The most common notation is 1.3 (iii), in which we use a single letter with double subscripts to designate the elements, the first subscript designating the row, and the second one the column, in which the elements appear.

Sometimes it is convenient to use letters for subscripts also, as shown in 1.4.

$$\begin{bmatrix} a_{11} & a_{12} & \cdots & a_{1m} \\ a_{21} & a_{22} & \cdots & a_{2m} \\ \cdot & \cdot & & \cdot \\ \cdot & \cdot & & \cdot \\ \cdot & \cdot & & \cdot \\ a_{n1} & a_{n2} & \cdots & a_{nm} \end{bmatrix} \quad [1.4]$$

This matrix has n rows and m columns; a_{ij} denotes its typical element. Since the exact numerical values of n and m are not specified, it is not possible to write the matrix in full. So we use dots (ellipses), to indicate the elements not written. Thus, in the first row the dots (ellipses) after a_{12} and before a_{1m} indicate that elements between them are not written. Similarly, the dots (ellipses) in column 1 indicate that elements between a_{21} and a_{n1} are not written. A notation closely resembling the one used in 1.4 is shown in 1.5:

$$\begin{bmatrix} a_{11} & \cdots & a_{1m} \\ \cdot & & \cdot \\ \cdot & & \cdot \\ \cdot & & \cdot \\ a_{n1} & \cdots & a_{nm} \end{bmatrix} \quad [1.5]$$

It is often informative to write

$$\mathbf{A}_{n\times m} = ((a_{ij}))$$

indicating that the matrix **A** with n rows and m columns has typical element a_{ij}. Thus $\mathbf{B}_{2\times3} = ((b_{ij}))$ means.

$$\mathbf{B} = \begin{bmatrix} b_{11} & b_{12} & b_{13} \\ b_{21} & b_{22} & b_{23} \end{bmatrix}$$

Matrices containing only one row are often called *row vectors*, and, similarly, matrices containing only one column are called *column vectors*. Thus $[a \quad b \quad c]$ is a row vector and $\begin{bmatrix} a \\ b \end{bmatrix}$ is a column vector. We shall return to vectors later.

Equality of Matrices

Two matrices are equal if (a) they both have the same number of rows and the same number of columns, and (b) their corresponding elements are equal. In symbols, if

$$\mathbf{A}_{n\times m} = ((a_{ij})) \quad \text{and} \quad \mathbf{B}_{r\times s} = ((b_{ij}))$$

then $\mathbf{A} = \mathbf{B}$, i.e., **A** and **B** are equal if $n = r$ and $m = s$ and $a_{ij} = b_{ij}$ for $i = 1,2, \ldots, n(=r)$; $j = 1,2, \ldots, m(=s)$.

$$\text{If} \quad \mathbf{A} = \begin{bmatrix} 1 & 2 \\ 3 & 4 \end{bmatrix} \quad \text{and} \quad \mathbf{B} = \begin{bmatrix} 1 & x \\ y & 4 \end{bmatrix}$$

then $\mathbf{A} = \mathbf{B}$ implies that $x = 2$ and $y = 3$. Similarly, if $\mathbf{M} = \mathbf{B}$ where

$$\mathbf{M} = \begin{bmatrix} p & 0 & 5 \\ q & 1 & 3 \end{bmatrix} \quad \text{and} \quad \mathbf{B} = \begin{bmatrix} 4 & 0 & 5 \\ 2 & 1 & 3 \end{bmatrix}$$

then $p = 4$ and $q = 2$. If

$$\mathbf{N} = \begin{bmatrix} 1 & 3 \\ 4 & 1 \end{bmatrix} \quad \text{and} \quad \mathbf{W} = \begin{bmatrix} a & b & c \\ d & e & f \end{bmatrix}$$

then **N** cannot be equal to **W** because they do not have the same number of columns.

Addition and Subtraction of Matrices

We define addition (subtraction) of matrices in terms of addition (subtraction) of their corresponding elements. The sum of two n by m matrices is an n by m matrix whose elements are the sum of the corresponding elements of the original matrices. In symbols, if $\mathbf{A}_{n\times m} = ((a_{ij}))$ and $\mathbf{B}_{n\times m} = ((b_{ij}))$, then their sum, denoted by $\mathbf{A} + \mathbf{B}$, is $((a_{ij} + b_{ij}))$. Thus, if

$$\mathbf{A} = \begin{bmatrix} a_{11} & a_{12} \\ a_{21} & a_{22} \end{bmatrix} \qquad \mathbf{B} = \begin{bmatrix} b_{11} & b_{12} \\ b_{21} & b_{22} \end{bmatrix}$$

then their sum is

$$\mathbf{A} + \mathbf{B} = \begin{bmatrix} (a_{11} + b_{11}) & (a_{12} + b_{12}) \\ (a_{21} + b_{21}) & (a_{22} + b_{22}) \end{bmatrix}$$

By way of numerical examples

$$\begin{bmatrix} 1 & 2 \\ 3 & 4 \end{bmatrix} + \begin{bmatrix} 5 & 6 \\ 7 & 8 \end{bmatrix} = \begin{bmatrix} (1+5) & (2+6) \\ (3+7) & (4+8) \end{bmatrix} = \begin{bmatrix} 6 & 8 \\ 10 & 12 \end{bmatrix}$$

$$\begin{bmatrix} 1 + \sqrt{3} & 0 \\ -1 & 1 - \sqrt{3} \end{bmatrix} + \begin{bmatrix} -\sqrt{3} & 0 \\ 1 & \sqrt{3} \end{bmatrix} = \begin{bmatrix} 1 & 0 \\ 0 & 1 \end{bmatrix}$$

$$\begin{bmatrix} 1 & 2 & 3 \\ 4 & 5 & 6 \\ 7 & 8 & 9 \end{bmatrix} + \begin{bmatrix} 0.1 & 0.2 & 0.3 \\ 0.4 & 0.5 & 0.6 \\ 0.7 & 0.8 & 0.9 \end{bmatrix} = \begin{bmatrix} 1.1 & 2.2 & 3.3 \\ 4.4 & 5.5 & 6.6 \\ 7.7 & 8.8 & 9.9 \end{bmatrix}$$

Note that we shall add two matrices only if they are of the same order. Addition is not defined for matrices of different orders. There is no question of adding a 3×4 matrix to a 4×5 matrix.

The definition of subtraction parallels the definition of addition.

$$\mathbf{A}_{n\times m} - \mathbf{B}_{n\times m} = ((a_{ij} - b_{ij}))$$

Thus, if

$$\mathbf{A} = \begin{bmatrix} -2 & 0 & 1 \\ 0 & 1 & -1 \end{bmatrix} \quad \text{and} \quad \mathbf{B} = \begin{bmatrix} 1 & -1 & 0 \\ 1 & 0 & -1 \end{bmatrix}$$

then

$$\mathbf{A} - \mathbf{B} = \begin{bmatrix} -2-1 & 0-(-1) & 1-0 \\ 0-1 & 1-0 & -1-(-1) \end{bmatrix} = \begin{bmatrix} -3 & 1 & 1 \\ -1 & 1 & 0 \end{bmatrix}.$$

Multiplication by a Scalar

Let k be an ordinary number (scalar) and $\mathbf{A} = ((a_{ij}))$ be any matrix. Then $k\mathbf{A} = ((ka_{ij}))$. That is, to multiply a matrix by an ordinary number (scalar), we multiply each element of the matrix by the number.

It is easy to verify that multiplication of a matrix by a positive integer is the same as repeated addition. Thus, $2\mathbf{Q} = \mathbf{Q} + \mathbf{Q}$, where $\mathbf{Q}$ is any matrix. Also, it is easy to see that

$$\mathbf{B}_{n \times m} - \mathbf{A}_{n \times m} = \mathbf{B}_{n \times m} + (-1)\mathbf{A}_{n \times m}$$

If $\mathbf{A} = \begin{bmatrix} 2 & 3 \\ 1 & 4 \end{bmatrix}$, twice $\mathbf{A} = \mathbf{A} + \mathbf{A} = \begin{bmatrix} 4 & 6 \\ 2 & 8 \end{bmatrix}$, and if $\mathbf{A} = \begin{bmatrix} 3 & 1 \\ 0 & -1 \end{bmatrix}$ and $\mathbf{B} = \begin{bmatrix} 0 & 1 \\ 1 & 0 \end{bmatrix}$, then (-1) times $\mathbf{B}$ is $\begin{bmatrix} 0 & -1 \\ -1 & 0 \end{bmatrix}$, and $\mathbf{A} - \mathbf{B} = \mathbf{A} + (-1)\mathbf{B} =$ $\begin{bmatrix} 3 & 1 \\ 0 & -1 \end{bmatrix} + \begin{bmatrix} 0 & -1 \\ -1 & 0 \end{bmatrix} = \begin{bmatrix} 3 & 0 \\ -1 & -1 \end{bmatrix}$.

Vectors

Obviously all that has been said so far about matrices apply to vectors, since vectors, as we mentioned earlier, are special kinds of matrices (having only a single column or a single row). Nonetheless, we shall repeat some of the main points with special reference to vectors. We shall denote a vector by a boldface, lower-case letter, and refer to a vector consisting of n elements by the term n-tuple.

Equality of two vectors. Two column vectors are said to be equal if they have the same number of elements and their corresponding elements are equal. Similarly, two row vectors are said to be equal if they have the same number of elements and their corresponding elements are equal. (In some books the term component is used instead of element.)

A row vector is never equal to a column vector. Equality of two vectors $\mathbf{a}$ and $\mathbf{b}$ is denoted by $\mathbf{a} = \mathbf{b}$.

Addition of vectors. The sum of two column vectors with the same number of elements (or of two row vectors with the same number of elements) is formed by adding the corresponding elements of the given vectors.

Addition of a row vector and a column vector is not defined. The sum of two vectors **a** and **b** is denoted by **a** + **b**. By way of illustration,

(1) $\begin{bmatrix} 9 \\ 3 \\ 4 \end{bmatrix} + \begin{bmatrix} 2 \\ 1 \\ 0 \end{bmatrix} = \begin{bmatrix} 11 \\ 4 \\ 4 \end{bmatrix}$;

(2) $[1 \quad 4 \quad -1] + [3 \quad 0 \quad 8] = [4 \quad 4 \quad 7]$;

(3) The sum of $\mathbf{c} = [1 \quad 2 \quad 3 \quad 4]$ and $\mathbf{d} = [3 \quad 5 \quad 6]$ is not defined, since **c** and **d** do not have the same number of elements;

(4) The sum of $\mathbf{a} = [1 \quad 2]$ and $\mathbf{b} = \begin{bmatrix} 2 \\ 1 \end{bmatrix}$ is not defined because **a** is a row vector, while **b** is a column vector.

Multiplication by a scalar. Let **a** be a column or row vector and k an ordinary number. Then the product $k\mathbf{a}$ is defined as the vector whose elements are k times the corresponding elements of **a**.

VECTOR REPRESENTATION OF A SYSTEM OF LINEAR EQUATIONS

The vector operations defined above can be used to express a system of linear equations compactly as a single vector equation. Consider, for example, the following two equations in two unknowns:

$$2x + 3y = 5$$
$$3x + 2y = 5$$

Let us form the following three column vectors, corresponding to the coefficients of x, those of y, and the constant terms:

$$\mathbf{a} = \begin{bmatrix} 2 \\ 3 \end{bmatrix}, \quad \mathbf{b} = \begin{bmatrix} 3 \\ 2 \end{bmatrix}, \quad \mathbf{c} = \begin{bmatrix} 5 \\ 5 \end{bmatrix}$$

Now the given set of equations can be expressed compactly as

$$x\mathbf{a} + y\mathbf{b} = \mathbf{c}$$

To check this out, we note that the vector equation just written is equivalent to

$$x \begin{bmatrix} 2 \\ 3 \end{bmatrix} + y \begin{bmatrix} 3 \\ 2 \end{bmatrix} = \begin{bmatrix} 5 \\ 5 \end{bmatrix}$$

which, by virtue of the definition of scalar multiplication, becomes

$$\begin{bmatrix} 2x \\ 3x \end{bmatrix} + \begin{bmatrix} 3y \\ 2y \end{bmatrix} = \begin{bmatrix} 5 \\ 5 \end{bmatrix}$$

which, in turn, by virtue of the definition of addition is the same as

$$\begin{bmatrix} (2x + 3y) \\ (3x + 2y) \end{bmatrix} = \begin{bmatrix} 5 \\ 5 \end{bmatrix}$$

and now the definition of equality gives

$$\begin{aligned} 2x + 3y &= 5 \\ 3x + 2y &= 5 \end{aligned}$$

INNER PRODUCTS

Let $\mathbf{a}'$ be a row vector and $\mathbf{b}$ a column vector, both being n-tuples, that is, vectors having n elements:

$$\mathbf{a}' = [a_1 \; . \; . \; . \; a_n]$$

$$\mathbf{b} = \begin{bmatrix} b_1 \\ \cdot \\ \cdot \\ \cdot \\ b_n \end{bmatrix}$$

Then the product $\mathbf{a}'$ times $\mathbf{b}$ is defined to be the scalar $a_1b_1 + \ldots + a_nb_n$. This product is denoted by $\mathbf{a'b}$ or $\mathbf{a}' \cdot \mathbf{b}$ or $\mathbf{a}' * \mathbf{b}$. It is sometimes called the *inner product* of $\mathbf{a}'$ and $\mathbf{b}$. To give an example,

$$[1 \quad 2 \quad 4] \begin{bmatrix} 2 \\ 4 \\ 3 \end{bmatrix} = (1)(2) + (2)(4) + (4)(3) = 22$$

Note that the inner product is defined for the particular circumstance in which the first (i.e., the left) vector is a row vector and the second (i.e., the right) vector is a column vector, both having the same number of elements.

By way of application, we shall now show that quantities such as the arithmetic mean of a set of numbers can be expressed as inner products. Suppose we are given the heights of five children in a day-care center:

Child number	1	2	3	4	5
Height	30	32	31	33	35

The sum of the heights can be expressed as the inner product

$$[1 \quad 1 \quad 1 \quad 1 \quad 1] \begin{bmatrix} 30 \\ 32 \\ 31 \\ 33 \\ 35 \end{bmatrix}$$

and the average height as the inner product

$$[1/5 \quad 1/5 \quad 1/5 \quad 1/5 \quad 1/5] \begin{bmatrix} 30 \\ 32 \\ 31 \\ 33 \\ 35 \end{bmatrix}$$

We also note that the sum of squares of the heights can be expressed as the inner product

$$[30 \quad 32 \quad 31 \quad 33 \quad 35] \begin{bmatrix} 30 \\ 32 \\ 31 \\ 33 \\ 35 \end{bmatrix}$$

This last inner product suggests that it is useful to define an operation called *transposition*, whereby the ith element of a given column vector is entered as the ith element of a corresponding row vector, and vice versa.

Thus, we say that [2   3] and $\begin{bmatrix} 2 \\ 3 \end{bmatrix}$ are *transposes* of each other. More generally, $[a_1 \ldots a_n]$ is the transpose of

$$\begin{bmatrix} a_1 \\ \cdot \\ \cdot \\ \cdot \\ a_n \end{bmatrix}$$

and vice versa. The usual notation for the transpose of $\mathbf{a}$ is $\mathbf{a}'$ or $\mathbf{a}^{\mathrm{T}}$. It is easy to see that the transpose of a transpose of a vector is the original vector. In symbols, $(\mathbf{a}')' = \mathbf{a}$. Thus, if $\mathbf{a} = \begin{bmatrix} 2 \\ 3 \end{bmatrix}$, then $\mathbf{a}' = [2 \quad 3]$, and $(\mathbf{a}')' = \begin{bmatrix} 2 \\ 3 \end{bmatrix}$. (Since any matrix can be viewed as a concatenation of vectors, the operation of transposition can be extended to matrices. Thus, if $\mathbf{A}$ is an $n \times m$ matrix, then the $m \times n$ matrix $\mathbf{A}'$ obtained by interchanging the rows and columns of $\mathbf{A}$ is called the *transpose* of $\mathbf{A}$. For example, $\begin{bmatrix} 3 & 8 \\ 4 & 1 \end{bmatrix}$ and $\begin{bmatrix} 3 & 4 \\ 8 & 1 \end{bmatrix}$ are transposes of each other.)

Using the operation of transposition of vectors, the following compact expression for sum of squares is obtained:

$$x_1^2 + \ldots + x_n^2 = \mathbf{x}'\mathbf{x}$$

where

$$\mathbf{x} = \begin{bmatrix} x_1 \\ \cdot \\ \cdot \\ \cdot \\ x_n \end{bmatrix} \quad \text{and } \mathbf{x}' = [x_1 \ldots x_n]$$

Similarly, given two sets of n ordered numbers $(x_1, \ldots, x_n)$ and $(y_1, \ldots, y_n)$, we have the following compact expression for their sum of products:

$$\mathbf{x}'\mathbf{y} = \mathbf{y}'\mathbf{x} = x_1 y_1 + \ldots + x_n y_n$$

where

$$\mathbf{x} = \begin{bmatrix} x_1 \\ \cdot \\ \cdot \\ \cdot \\ x_n \end{bmatrix} \qquad \mathbf{y} = \begin{bmatrix} y_1 \\ \cdot \\ \cdot \\ \cdot \\ y_n \end{bmatrix}$$

Matrix-Vector Multiplication

Consider the frequency distribution of localities (places) by population size, given in Table 1.3. It is easy to verify that the cumulative frequencies shown in column 3 of Table 1.3 correspond to the following inner products:

$$[1 \quad 1 \quad 1 \quad 1] \begin{bmatrix} 1 \\ 4 \\ 10 \\ 25 \end{bmatrix} = 40$$

$$[0 \quad 1 \quad 1 \quad 1] \begin{bmatrix} 1 \\ 4 \\ 10 \\ 25 \end{bmatrix} = 39$$

$$[0 \quad 0 \quad 1 \quad 1] \begin{bmatrix} 1 \\ 4 \\ 10 \\ 25 \end{bmatrix} = 35$$

$$[0 \quad 0 \quad 0 \quad 1] \begin{bmatrix} 1 \\ 4 \\ 10 \\ 25 \end{bmatrix} = 25$$

If we stack up these inner products we get the following result:

$$\begin{bmatrix} 1 & 1 & 1 & 1 \\ 0 & 1 & 1 & 1 \\ 0 & 0 & 1 & 1 \\ 0 & 0 & 0 & 1 \end{bmatrix} \begin{bmatrix} 1 \\ 4 \\ 10 \\ 25 \end{bmatrix} = \begin{bmatrix} 40 \\ 39 \\ 35 \\ 25 \end{bmatrix}$$

which suggests the following definition of matrix-vector multiplication.

Let A be a matrix and v a column vector such that the number of columns of A equals the number of elements in v. Then the product A times v, written Av, or A·v or A*v, is a column vector c whose i^{th} element is equal to the inner product of the ith row of A with v.

Thus in the product

$$\begin{bmatrix} 0 & 0 & 0 & 1 \\ 0 & 0 & 1 & 1 \\ 0 & 1 & 1 & 1 \\ 1 & 1 & 1 & 1 \end{bmatrix} \begin{bmatrix} 1 \\ 4 \\ 10 \\ 25 \end{bmatrix} = \mathbf{c}$$

the first element of **c** is $[0 \quad 0 \quad 0 \quad 1] \begin{bmatrix} 1 \\ 4 \\ 10 \\ 25 \end{bmatrix}$, i.e., 25; the second element is $[0 \quad 0 \quad 1 \quad 1] \begin{bmatrix} 1 \\ 4 \\ 10 \\ 25 \end{bmatrix}$, i.e., 35; the third element is $[0 \quad 1 \quad 1 \quad 1] \begin{bmatrix} 1 \\ 4 \\ 10 \\ 25 \end{bmatrix}$, i.e., 39; the fourth element, $[1 \quad 1 \quad 1 \quad 1] \begin{bmatrix} 1 \\ 4 \\ 10 \\ 25 \end{bmatrix}$, i.e., 40.

Similarly,

$$\begin{bmatrix} 1 & 0 & 0 & 0 \\ 1 & 1 & 0 & 0 \\ 1 & 1 & 1 & 0 \\ 1 & 1 & 1 & 1 \end{bmatrix} \begin{bmatrix} 1 \\ 4 \\ 10 \\ 25 \end{bmatrix} = \begin{bmatrix} 1 \\ 5 \\ 15 \\ 40 \end{bmatrix}$$

which, incidentally, illustrates the calculation of cumulatives from the top.

NOTE:

(1) If A is a 2 × 3 matrix, and Av is defined (where v is a column vector), then we know that v has 3 elements.

TABLE 1.3

Population Size Class	*Number of Localities*	*Cumulative from Bottom*	*Cumulative from Top*
50 or more	1	40	1
20-49	4	39	5
5-19	10	35	15
Below 5	25	25	40

(2) **If C is a 2 × 3 matrix, and u is a column vector with 2 elements only, then Cu is not defined.**

(3) If $A_{n\times m} = ((a_{ij}))$ and $\mathbf{v} = \begin{bmatrix} v_1 \\ \cdot \\ \cdot \\ \cdot \\ v_k \end{bmatrix}$, then

Av = c implies that $k = m$ and that c has n elements.

(4) **The product of a *row* vector (on the left) times a matrix (on the right) may be defined as follows: Let u′ be a row vector with n elements and let A be any $n \times m$ matrix. Then the product u′ (on the left) times A (on the right), denoted as u′A or u′·A or u′*A, is a row vector whose j^{th} element equals the inner product of u′ with the jth column of A. Thus, for example,**

$$[1 \quad -1] \begin{bmatrix} 1 & 2 & 3 \\ 4 & 5 & 6 \end{bmatrix} = [-3 \quad -3 \quad -3]$$

Incidentally, this particular example illustrates an operation which gives the difference: the first row of a 2 by 3 matrix *minus* its second row. The following example illustrates an operation that yields the sum of the two rows of a 2 × 3 matrix:

$$[1 \quad 1] \begin{bmatrix} a & c & e \\ b & d & f \end{bmatrix} = [(a+b) \; (c+d) \; (e+f)]$$

(5) **If A is an $n \times m$ matrix, u′ is a $1 \times r$ vector and d′ a $1 \times k$ vector, then u′A = d′ implies $r = n$ and $k = m$. Also, the jth element of d′ equals the inner product of u′ with the jth column of A.**

Matrix Multiplication

Suppose a department has a two-track graduate program, emphasizing, say, "basic" research and "application," respectively. Let us assume that a typical student in the basic program takes four required courses and eight electives, whle one in the applied program takes two required courses and twelve electives. This structure can be depicted in matrix form as shown below:

	Number of required courses	Number of electives	
A =	4	8	Basic
	2	12	Applied

$$\mathbf{A} = \begin{bmatrix} 4 & 8 \\ 2 & 12 \end{bmatrix}$$

Suppose, on the average, in the required courses a student devotes a_1 hours to directed study and a_2 hours to independent study, per course, while the corresponding figures in electives are b_1 and b_2, respectively. Let us depict this feature also in matrix form:

	Directed Study	Independent Study	
B =	a_1	a_2	Required
	b_1	b_2	Elective

$$\mathbf{B} = \begin{bmatrix} a_1 & a_2 \\ b_1 & b_2 \end{bmatrix}$$

Now, suppose based on these pieces of "information" we are asked to calculate the hours devoted to directed study and independent study, respectively, by a typical student in the basic program and by one in the applied program. It is easy to see that the figures we are looking for are those in the following matrix:

	Directed Study	Independent Study	
C =	$4a_1 + 8b_1$	$4a_2 + 8b_2$	Basic
	$2a_1 + 12b_1$	$2a_2 + 12b_2$	Applied

$$\mathbf{C} = \begin{bmatrix} 4a_1 + 8b_1 & 4a_2 + 8b_2 \\ 2a_1 + 12b_1 & 2a_2 + 12b_2 \end{bmatrix}$$

Notice that the first column of **C** is the matrix-vector product **A** times the first column of **B**, while the second column of **C** is **A** times the second column of **B**. This example suggests the usefulness of defining an operation that produces a matrix (**C** in our example) by concatenating horizontally a given matrix (**A** in our example) times the successive columns of another (**B** in the example). We define such a concatenation involving **A** and **B** the product **A** times **B**, usually denoted by **AB**, or **A·B** or **A*B**. The operation that produces such a concatenation is called *matrix-matrix multiplication* or simply *matrix multiplication*. Using the matrices introduced above, we say that

$$\mathbf{AB} = \mathbf{C}$$

stipulating that, as mentioned above, **AB** means the horizontal concatenation in which **A** times the first column of **B** is followed on the right by **A** times the second column of **B**.

Notice that this operation (i.e., matrix multiplication as defined above) applies only if the number of columns in the left-factor (**A** in our example) equals the number of rows in the right-factor (**B** in the example). Obviously, if this condition does not prevail, the matrix-vector product **A** times j^{th} column of **B** is not defined, for any j. For this reason if

$$\mathbf{E} = \begin{bmatrix} a & b \\ c & d \end{bmatrix} \quad \text{and} \quad \mathbf{F} = \begin{bmatrix} e & f & g \\ h & i & j \end{bmatrix}$$

then the product **EF** is defined, while **FE** is not.

Let us now illustrate the important point that matrix multiplication, in general, is not commutative, i.e., that if the factors are rearranged, the product may be altered. Consider the matrices

$$\mathbf{M} = \begin{bmatrix} 1 & 2 \\ 3 & 4 \end{bmatrix} \quad \text{and} \quad \mathbf{N} = \begin{bmatrix} 5 & 6 \\ 7 & 8 \end{bmatrix}$$

In this case both **MN** and **NM** are defined. But

$$\mathbf{MN} = \left[\begin{bmatrix} 1 & 2 \\ 3 & 4 \end{bmatrix} \begin{bmatrix} 5 \\ 7 \end{bmatrix} \ \begin{bmatrix} 1 & 2 \\ 3 & 4 \end{bmatrix} \begin{bmatrix} 6 \\ 8 \end{bmatrix} \right] = \begin{bmatrix} 19 & 22 \\ 43 & 50 \end{bmatrix}$$

while

$$\mathbf{NM} = \left[\begin{bmatrix} 5 & 6 \\ 7 & 8 \end{bmatrix} \begin{bmatrix} 1 \\ 3 \end{bmatrix} \ \begin{bmatrix} 5 & 6 \\ 7 & 8 \end{bmatrix} \begin{bmatrix} 2 \\ 4 \end{bmatrix} \right] = \begin{bmatrix} 23 & 34 \\ 31 & 46 \end{bmatrix}$$

demonstrating that the product is altered when the factors are rearranged.

Because matrix multiplication is not, in general, commutative, we give special attention to the order of factors and speak of the product **AB** as the result of premultiplying **B** by **A** or postmultiplying **A** by **B**.

NOTE:

(1) An alternative way of defining matrix multiplication is the following: Given $\mathbf{A}_{n \times m} = ((a_{ij}))$ and $\mathbf{B}_{m \times p} = ((b_{ij}))$, the product **AB** is an $(n \times p)$ matrix whose (i, j) element equals the inner product of the ith row of **A** with the jth column of **B**. Thus

$$.144 \begin{bmatrix} 1 & 2 \\ 3 & 4 \end{bmatrix} \begin{bmatrix} 5 & 6 \\ 7 & 8 \end{bmatrix} = \begin{bmatrix} [1 \quad 2] \begin{bmatrix} 5 \\ 7 \end{bmatrix} & [1 \quad 2] \begin{bmatrix} 6 \\ 8 \end{bmatrix} \\ [3 \quad 4] \begin{bmatrix} 5 \\ 7 \end{bmatrix} & [3 \quad 4] \begin{bmatrix} 6 \\ 8 \end{bmatrix} \end{bmatrix} = \begin{bmatrix} 19 & 22 \\ 43 & 50 \end{bmatrix}$$

(2) Even when the product **AB** exists, the product **BA** may not be defined. This is the case, for example, if **A** is of order (2×3) while **B** is of order (3×7).

(3) If the product **AB** exists, the factors **A** and **B** are said to be *conformable* (with respect to multiplication).

(4) If the factors of **AB** are conformable (i.e., if the number of columns in **A** is equal to the number of rows in **B**), and the factors of **BC** are conformable (i.e., if the number of columns in **B** is equal to the number of rows in **C**), then the product **ABC** exists and is equal to **A**(**BC**) and to (**AB**)**C**, where parentheses signify the operation that has priority (the one that should be performed first). Put differently, assuming that the matrices are conformable, we can get the product **ABC** either by postmultiplying **AB** by **C** or premultiplying **BC** by **A**.

(5) Assuming that the factors of **AB** are conformable and that those of **AC** are also conformable, the product **A**(**B** + **C**), where parentheses signify operation that has priority, is equivalent to **AB** + **AC**. Similarly, given conformability, (**E** + **F**)**G** = **EG** + **FG**.

EXAMPLES OF THE USE OF MATRIX MULTIPLICATION

Matrices in Regression Analysis. Consider the following pairs of observations on Y = weight and X = height:

Child number	1	2	3	4	5	6	7	8	9
Y: weight	63	70	51	62	54	60	75	58	53
X: height	57	58	47	57	50	52	53	49	40

Suppose we hypothesize that there is a straight-line regression relationship between weight and height. Under this hypothesis we write the following equations connecting the observed values of X and Y:

$$63 = \beta_0 + 57\beta_1 + e_1$$
$$70 = \beta_0 + 58\beta_1 + e_2$$

and so on, where $e_1, e_2, \ldots$ are residuals, These equations can be compactly written using matrices. We define three vectors and one matrix

$$\mathbf{y} = \begin{bmatrix} 63 \\ 70 \\ 51 \\ 62 \\ 54 \\ 60 \\ 75 \\ 58 \\ 53 \end{bmatrix} \quad \mathbf{X} = \begin{bmatrix} 1 & 57 \\ 1 & 58 \\ 1 & 47 \\ 1 & 57 \\ 1 & 50 \\ 1 & 52 \\ 1 & 53 \\ 1 & 49 \\ 1 & 40 \end{bmatrix} \quad \boldsymbol{\beta} = \begin{bmatrix} \beta_0 \\ \beta_1 \end{bmatrix} \quad \mathbf{e} = \begin{bmatrix} e_1 \\ e_2 \\ e_3 \\ e_4 \\ e_5 \\ e_6 \\ e_7 \\ e_8 \\ e_9 \end{bmatrix}$$

The observational equations can now be written as

$$\mathbf{y}_{9\times1} = \mathbf{X}_{9\times2}\boldsymbol{\beta}_{2\times1} + \mathbf{e}_{9\times1}$$

Notice that the first column of the **X** matrix contains only 1's. This is true of all regression models in which there is a constant term β_0.

Markov chains. As another area of application, let us consider Markov chains. A Markov chain is a chance process having the special property that one can predict its future just as accurately from a knowledge of the present state of affairs as from a knowledge of the present as well as the entire past history. By way of illustration, consider intergenerational mobility in a society. For simplicity, let us imagine that there are only two social classes, say, the upper and the lower. Considering only the eldest sons, we say that the mobility process is a Markov process if the probability that a man is in a particular class is dependent only on the class of his father, but not further dependent on the class of his grandfather or of any of the more distant members of the male line.

The probability that a man is in a particular class (e.g., lower class), given that his father is in a particular class (say, upper class) is known as

TABLE 1.4
Transition Probabilities: Intergenerational Mobility Process

Father's Social Class	Son's Social Class	
	Upper	*Lower*
Upper	0.4	0.6
Lower	0.1	0.9

a transition probability. With two classes, a 2×2 matrix of transition probabilities, as illustrated in Table 1.4, determines the fundamental properties of a Markov chain.

According to these figures, the chance that an upper-class father's son (remember that we are considering only the eldest sons) moves down to the lower class is 0.6, and similarly the probability of upward mobility for those in the lower class is 0.1.

Suppose at the beginning of the observation period 20 percent of the (initial) generation are in the upper class and the rest in the lower class. If there are N persons in the initial generation, then $0.2N$ are in the upper class and $0.8N$ in the lower class. According to the transition probabilities given above, we would expect $(0.2N)(0.4) + (0.8N)(0.1)$ of the next generation to be in the upper class and $(0.2N)(0.6) + (0.8N)(0.9)$ in the lower class. The class structure of the second generation is thus expected to be

$$[(0.2)N \quad (0.8)N] \begin{bmatrix} 0.4 & 0.6 \\ 0.1 & 0.9 \end{bmatrix}$$

which is the initial class structure postmultiplied by the transition probability matrix. To get the class structure of the third generation we similarly postmultiply the class structure of the second generation by the transition probability matrix. If the transition probabilities remain unchanged, the class structure of the third generation is obtained as

$$[(0.2)N \quad (0.8)N] \begin{bmatrix} 0.4 & 0.6 \\ 0.1 & 0.9 \end{bmatrix} \begin{bmatrix} 0.4 & 0.6 \\ 0.1 & 0.9 \end{bmatrix}$$

which can be written as

$$[(0.2)N \quad (0.8)N] \begin{bmatrix} 0.4 & 0.6 \\ 0.1 & 0.9 \end{bmatrix}^2$$

following the familiar convention in ordinary algebra of writing x^2 for x times x.

Similarly, if the transition probabilities remain unchanged, the expected class structure of the fourth generation is

$$[(0.2)N \quad (0.8)N] \begin{bmatrix} 0.4 & 0.6 \\ 0.1 & 0.9 \end{bmatrix}^3$$

and so on. If you try to derive these results (the class structure of successive generations) without using matrices, you would find your task extremely cumbersome.

An interesting property of transition probability matrices of the kind introduced above is that for sufficiently large values of n, their powers higher than n are indistinguishable. Thus,

$$\begin{bmatrix} 0.4 & 0.6 \\ 0.1 & 0.9 \end{bmatrix}^4 = \begin{bmatrix} 0.1498 & 0.8502 \\ 0.1417 & 0.8583 \end{bmatrix}$$

$$\begin{bmatrix} 0.4 & 0.6 \\ 0.1 & 0.9 \end{bmatrix}^8 = \begin{bmatrix} 0.14291338 & 0.85708662 \\ 0.14284777 & 0.85715223 \end{bmatrix}$$

indicating that for large n, powers of

$$\begin{bmatrix} 0.4 & 0.6 \\ 0.1 & 0.9 \end{bmatrix}$$

higher than n are indistinguishable from

$$\begin{bmatrix} 1/7 & 6/7 \\ 1/7 & 6/7 \end{bmatrix}$$

We return to this topic in Chapter 4.

THE IDENTITY MATRIX

In ordinary algebra we have the number 1, which has the property that its product with any number is the number itself. We now introduce an analogous concept in matrix algebra.

Consider the following 2×2 matrices:

$$\mathbf{A} = \begin{bmatrix} 5 & 2 \\ 3 & 4 \end{bmatrix} \quad \text{and} \quad \mathbf{B} = \begin{bmatrix} 1 & 0 \\ 0 & 1 \end{bmatrix}$$

The product **A** times **B** and the product **B** times **A** in this case are both equal to **A** as can be easily verified. That is, **AB** = **BA** = **A**. In the algebra of 2 by 2 matrices

$$\begin{bmatrix} 1 & 0 \\ 0 & 1 \end{bmatrix}$$

is called the *identity* matrix under multiplication. (It behaves like unity in ordinary algebra.) It is a *square* matrix—one having the same number of rows as columns—and it has unity in the *principal diagonal* (i.e., the diagonal of elements from the upper left corner to the lower right corner) and 0 everywhere else.

There are identity matrices of different orders. Thus,

$$\begin{bmatrix} 1 & 0 \\ 0 & 1 \end{bmatrix} \begin{bmatrix} 1 & 0 & 0 \\ 0 & 1 & 0 \\ 0 & 0 & 1 \end{bmatrix} \begin{bmatrix} 1 & 0 & 0 & 0 \\ 0 & 1 & 0 & 0 \\ 0 & 0 & 1 & 0 \\ 0 & 0 & 0 & 1 \end{bmatrix}$$

are identity matrices of orders 2×2, 3×3, and 4×4, respectively. In general, an $n \times n$ identity matrix has its (1,1), 2,2), . . . , (n,n) elements each equal to 1 and all other elements equal to 0. The usual notation for an $n \times n$ identity matrix is $\mathbf{I}_{n \times n}$ or $\mathbf{I}_n$ or **I**, the last when the order is obvious.

2. ELEMENTARY OPERATIONS AND THE INVERSE OF A MATRIX

This chapter is concerned with some further properties of matrices and the matrix as an operator. The main objective is to introduce the concept "inverse of a matrix," and to illustrate its applications.

Elementary Operations

The first concept we shall introduce is that of elementary operations. To see why such operations are of interest, let us consider the following two equations in two unknowns:

$$\begin{aligned} 2x + 3y &= 5 \\ 3x - 6y &= -3 \end{aligned} \qquad [2.1]$$

To solve this set of equations, let us adopt the elementary approach consisting of elimination of one of the two unknowns to get an equation in one unknown. Multiplication of the first equation by 2 gives

$$4x + 6y = 10 \qquad [2.2]$$

Adding this to the second equation in the initial set, we get

$$7x = 7 \qquad [2.3]$$

leading to the solution $x = 1$; $y = 1$.

The first two steps just carried out—namely (1) multiplying an equation by a scalar, and (2) adding one equation to another—can be performed by matrix multiplication, as illustrated below.

Let us write the initial set of equations (2.1) in matrix form:

$$\begin{bmatrix} 2 & 3 \\ 3 & -6 \end{bmatrix} \begin{bmatrix} x \\ y \end{bmatrix} = \begin{bmatrix} 5 \\ -3 \end{bmatrix} \qquad [2.4]$$

If we premultiply both sides of this matrix equation by

$$\mathbf{E}_1 = \begin{bmatrix} 2 & 0 \\ 0 & 1 \end{bmatrix}$$

we get

$$\begin{bmatrix} 4 & 6 \\ 3 & -6 \end{bmatrix} \begin{bmatrix} x \\ y \end{bmatrix} = \begin{bmatrix} 10 \\ -3 \end{bmatrix} \qquad [2.5]$$

which is equivalent to 2.2 paired with the second equation in 2.1. Now let us premultiply 2.5 by

$$\mathbf{E}_2 = \begin{bmatrix} 1 & 1 \\ 0 & 1 \end{bmatrix}$$

The result is

$$\begin{bmatrix} 7 & 0 \\ 3 & -6 \end{bmatrix} \begin{bmatrix} x \\ y \end{bmatrix} = \begin{bmatrix} 7 \\ -3 \end{bmatrix} \qquad [2.6]$$

which gives the equation $7x = 7$, obtained earlier, and the second equation of the initial set.

The matrices $\mathbf{E}_1$ and $\mathbf{E}_2$ used above are examples of what are known as *elementary operators*. (We encountered such operators in Chapter 1. See, for example, the matrices used for obtaining cumulative sums.)

Formally, there are three types of elementary row operations that may be carried out on a matrix:

(1) interchanging two rows,
(2) multiplying each element of a row by a nonzero scalar, and
(3) adding a nonzero multiple of one row to another.

Each of these operations on the rows of a matrix can be carried out by premultiplying the given matrix by an appropriate *elementary row operator*. To get the appropriate elementary row operator, all that we have to do is carry out the required operations on an $n \times n$ identity matrix, if the given matrix is of order $n \times m$. To illustrate, suppose we are given

$$\mathbf{A} = \begin{bmatrix} 2 & 3 \\ 1 & 0 \\ 1 & 1 \end{bmatrix}$$

To interchange rows 1 and 3, we premultiply **A** by

$$\mathbf{E}_1 = \begin{bmatrix} 0 & 0 & 1 \\ 0 & 1 & 0 \\ 1 & 0 & 0 \end{bmatrix}$$

which is obtained by interchanging the first and third rows of $\mathbf{I}_{3\times 3}$. Clearly,

$$\mathbf{E}_1\mathbf{A} = \begin{bmatrix} 0 & 0 & 1 \\ 0 & 1 & 0 \\ 1 & 0 & 0 \end{bmatrix} \begin{bmatrix} 2 & 3 \\ 1 & 0 \\ 1 & 1 \end{bmatrix} = \begin{bmatrix} 1 & 1 \\ 1 & 0 \\ 2 & 3 \end{bmatrix}$$

Notice that by premultiplying **A** by $\mathbf{E}_1$ we interchanged the first and third rows of **A**. Now suppose we want to multiply the second row of the same **A** by a scalar, say, (–8). The appropriate elementary row operator for this is

$$\mathbf{E}_2 = \begin{bmatrix} 1 & 0 & 0 \\ 0 & -8 & 0 \\ 0 & 0 & 1 \end{bmatrix}$$

which, it may be noted, is constructed by performing the required operation [multiplying the second row by (–8)] on a 3×3 identity matrix. As can be verified,

$$\mathbf{E}_2\mathbf{A} = \begin{bmatrix} 2 & 3 \\ -8 & 0 \\ 1 & 1 \end{bmatrix}$$

which is **A** with its second row multiplied by (–8), as desired.

If we wish to add twice the second row of **A** to the first, we perform the operation on a 3 × 3 identity matrix and use the resulting matrix as an elementary row operator:

$$\mathbf{E}_3 = \begin{bmatrix} 1 & 2 & 0 \\ 0 & 1 & 0 \\ 0 & 0 & 1 \end{bmatrix}$$

$$\mathbf{E}_3\mathbf{A} = \begin{bmatrix} 4 & 3 \\ 1 & 0 \\ 1 & 1 \end{bmatrix}$$

Elementary column operations can be defined similarly. They are equivalent to *postmultiplication* by appropriate *elementary column operators* (of order $m \times m$, if **A** is of order $n \times m$) which can be constructed by carrying out the specified elementary column operations on the identity matrix of the appropriate order. Two examples follow.

1. If **A** is the same matrix as the one given above, interchange of the first and second column can be achieved by postmultiplication of **A** by $\begin{bmatrix} 0 & 1 \\ 1 & 0 \end{bmatrix}$:

$$\begin{bmatrix} 2 & 3 \\ 1 & 0 \\ 1 & 1 \end{bmatrix} \begin{bmatrix} 0 & 1 \\ 1 & 0 \end{bmatrix} = \begin{bmatrix} 3 & 2 \\ 0 & 1 \\ 1 & 1 \end{bmatrix}$$

2. With the same matrix **A**, the addition of the first column to the second can be effected with postmultiplication of **A** by $\begin{bmatrix} 1 & 1 \\ 0 & 1 \end{bmatrix}$:

$$\begin{bmatrix} 2 & 3 \\ 1 & 0 \\ 1 & 1 \end{bmatrix} \begin{bmatrix} 1 & 1 \\ 0 & 1 \end{bmatrix} = \begin{bmatrix} 2 & 5 \\ 1 & 1 \\ 1 & 2 \end{bmatrix}$$

Echelon Matrices

Consider the set of equations

$$
\begin{aligned}
x + 2y + 3z &= 2 \\
x + 2y + 4z &= 1 \\
2x + 4y + 7z &= 3
\end{aligned}
\qquad [2.7]
$$

The coefficients of the equations form the matrix

$$\mathbf{B} = \begin{bmatrix} 1 & 2 & 3 \\ 1 & 2 & 4 \\ 2 & 4 & 7 \end{bmatrix}$$

Consider the following sequence of elementary row operations on **B**:

(1) Subtract row 1 from row 2.
(2) Subtract twice row 1 of the resulting matrix from its row 3.
(3) Subtract the new row 2 from the new row 3.

To perform these operations, the appropriate elementary row operators are

$$\mathbf{E}_1 = \begin{bmatrix} 1 & 0 & 0 \\ -1 & 1 & 0 \\ 0 & 0 & 1 \end{bmatrix} \quad \text{for operation (1),}$$

$$\mathbf{E}_2 = \begin{bmatrix} 1 & 0 & 0 \\ 0 & 1 & 0 \\ -2 & 0 & 1 \end{bmatrix} \quad \text{for operation (2),}$$

$$\mathbf{E}_3 = \begin{bmatrix} 1 & 0 & 0 \\ 0 & 1 & 0 \\ 0 & -1 & 1 \end{bmatrix} \quad \text{for operation (3).}$$

Applying these in sequence, we get

$$\mathbf{E}_1\mathbf{B} = \begin{bmatrix} 1 & 2 & 3 \\ 0 & 0 & 1 \\ 2 & 4 & 7 \end{bmatrix},$$

$$\mathbf{E}_2(\mathbf{E}_1\mathbf{B}) = \begin{bmatrix} 1 & 2 & 3 \\ 0 & 0 & 1 \\ 0 & 0 & 1 \end{bmatrix},$$

and

$$\mathbf{E}_3(\mathbf{E}_2\mathbf{E}_1\mathbf{B}) = \begin{bmatrix} 1 & 2 & 3 \\ 0 & 0 & 1 \\ 0 & 0 & 0 \end{bmatrix}$$

This last matrix illustrates the concept of an *echelon matrix*, which is any $n \times m$ matrix with the following properties:

(1) each of the first k rows ($0 \leq k \leq n$) has one or more nonzero elements;

(2) for each such row, the first nonzero element, when reading from left to right, is unity;

(3) the arrangement of these rows is such that the first nonzero element in the ith row appears in a column to the right of the column in which the first nonzero element appears in row $(i - 1)$;

(4) after the first k rows, the elements of the remaining rows (if any) are all zero.

Thus

$$\begin{bmatrix} 0 & 0 & 1 & -2 & 3 \\ 0 & 0 & 0 & 1 & 2 \\ 0 & 0 & 0 & 0 & 0 \end{bmatrix} \text{ and } \begin{bmatrix} 1 & 2 & 0 & 5 \\ 0 & 1 & -2 & 1 \\ 0 & 0 & 0 & 1 \end{bmatrix}$$

are echelon matrices.

For any given matrix $\mathbf{A}$, it is possible to find a sequence of elementary row operations that transforms $\mathbf{A}$ into an echelon matrix. To illustrate, consider the matrix $\mathbf{A}_0$ in 2.8:

$$\underset{\mathbf{A}_0}{\begin{bmatrix} 0 & 1 & 1 \\ 1 & 3 & -1 \\ 1 & 5 & 2 \end{bmatrix}}, \quad \underset{\mathbf{A}_1}{\begin{bmatrix} 1 & 3 & -1 \\ 0 & 1 & 1 \\ 1 & 5 & 2 \end{bmatrix}}, \quad \underset{\mathbf{A}_2}{\begin{bmatrix} 1 & 3 & -1 \\ 0 & 1 & 1 \\ 0 & 2 & 3 \end{bmatrix}}, \quad \underset{\mathbf{A}_3}{\begin{bmatrix} 1 & 3 & -1 \\ 0 & 1 & 1 \\ 0 & 0 & 1 \end{bmatrix}} \qquad [2.8]$$

The characteristic features of echelon matrices suggest that to transform $\mathbf{A}_0$ to an echelon form we may start by interchanging rows 1 and 2, thus getting $\mathbf{A}_1$ in 2.8. (Rows 1 and 2 of $\mathbf{A}_1$ satisfy the characteristic features of the first two rows of echelon matrices.) If we now subtract the first row of $\mathbf{A}_1$ from its third row we get $\mathbf{A}_2$ in 2.8. Subtraction of twice the second row of $\mathbf{A}_2$ from its third row gives $\mathbf{A}_3$, which is in echelon form.

Two points should be noted. First, the echelon matrix in *no* sense is "equal" to the given matrix from which it is derived. It is simply the result of a sequence of elementary row operations on the given matrix. Second, corresponding to a given matrix, there may be two or more echelon matrices. To illustrate this latter point, consider the matrix $\mathbf{A}_0$ in 2.8. Suppose we interchange the first row and third row (see $\mathbf{B}_1$ in 2.9); then subtract the new first row from the second ($\mathbf{B}_2$ in 2.9); divide the second by (–2), i.e., multiply it by (–½), ($\mathbf{B}_3$ in 2.9); and finally subtract the new second row from the third, and, after subtraction, multiply the resulting row by (–2) ($\mathbf{B}_4$ in 2.9); the result is an echelon matrix, which is not the same as $\mathbf{A}_3$ in 2.8.

$$\underset{\mathbf{A}_0}{\begin{bmatrix} 0 & 1 & 1 \\ 1 & 3 & -1 \\ 1 & 5 & 2 \end{bmatrix}}, \underset{\mathbf{B}_1}{\begin{bmatrix} 1 & 5 & 2 \\ 1 & 3 & -1 \\ 0 & 1 & 1 \end{bmatrix}}, \underset{\mathbf{B}_2}{\begin{bmatrix} 1 & 5 & 2 \\ 0 & -2 & -3 \\ 0 & 1 & 1 \end{bmatrix}}, \underset{\mathbf{B}_3}{\begin{bmatrix} 1 & 5 & 2 \\ 0 & 1 & 3/2 \\ 0 & 1 & 1 \end{bmatrix}}, \underset{\mathbf{B}_4}{\begin{bmatrix} 1 & 5 & 2 \\ 0 & 1 & 3/2 \\ 0 & 0 & 1 \end{bmatrix}} \quad [2.9]$$

The Inverse of a Square Matrix

The matrix operations of addition, subtraction, and multiplication were introduced in Chapter 1. In arithmetic and ordinary algebra, however, there is also an operation of division. Can we define an analogous operation for matrices? Strictly speaking, there is no such thing as division of one matrix by another; but there is an operation that accomplishes the same thing as division does in arithmetic and scalar algebra.

In arithmetic, we know that multiplying by 2^{-1} is the same thing as dividing by 2. More generally, given any nonzero scalar a, we can speak of multiplying by a^{-1} instead of dividing by a. The multiplication by a^{-1} has the property that

$$aa^{-1} = a^{-1}a = 1 \quad [2.10]$$

This prompts the question, for a matrix $\mathbf{A}$, can we find a matrix $\mathbf{B}$ such that

$$\mathbf{BA} = \mathbf{AB} = \mathbf{I}_{n \times n} \quad [2.11]$$

where $\mathbf{I}$ is an identity matrix of order n (the matrix analogue of unity referred to in 2.10).

In order for 2.11 to hold, **AB** and **BA** must be of order $n \times n$; but **AB** is of order $n \times n$ only if **A** has n rows and **B** has n columns, and **BA** is of order $n \times n$ only of **B** has n rows and **A** has n columns. Therefore, 2.11 holds only if **A** and **B** are both of order $n \times n$. (As mentioned earlier, matrices having as many rows as columns are called *square* matrices.) This leads to the following definition.

Given a square matrix **A**, if there exists a square matrix **B**, such that

$$\mathbf{BA} = \mathbf{AB} = \mathbf{I} \qquad [2.12]$$

then **B** is called the *inverse matrix* (or simply the *inverse*) of **A**, and **A** is said to be *invertible*. Not all square matrices are invertible, as we will see later. (A square matrix that does *not* have an inverse is said to be *singular*. A square matrix that possesses an inverse is said to be *nonsingular*.)

To illustrate the concept of inverse, given a matrix

$$\mathbf{A} = \begin{bmatrix} 1 & 1 \\ 3 & 4 \end{bmatrix}$$

it is easy to verify that the matrix

$$\mathbf{B} = \begin{bmatrix} 4 & -1 \\ -3 & 1 \end{bmatrix}$$

satisfies the relations

$$\mathbf{AB} = \mathbf{BA} = \mathbf{I}$$

Hence $\begin{bmatrix} 4 & -1 \\ -3 & 1 \end{bmatrix}$ is the inverse of **A**. Similarly, given

$$\mathbf{C} = \begin{bmatrix} -2 & 0 & 1 \\ -3 & 1 & 4 \\ -5 & 0 & 2 \end{bmatrix}$$

the matrix

$$\mathbf{D} = \begin{bmatrix} 2 & 0 & -1 \\ -14 & 1 & 5 \\ 5 & 0 & -2 \end{bmatrix}$$

satisfies the relations

$$\mathbf{CD} = \mathbf{DC} = \mathbf{I}$$

Hence **D** is the inverse of **C**.

A Procedure to Calculate the Inverse of a Matrix If It Exists

Let us first confine attention to finding a matrix **B**, given a matrix **A**, such that the premultiplication requirement **BA** = **I** is satisfied. (For the moment, we are *not* concerned with the postmultiplication requirement that **AB** should also be equal to **I**.) Our task, then, is to find a premultiplying matrix $\mathbf{B}_{n \times n}$ that transforms $\mathbf{A}_{n \times n}$ into the identity matrix **I**. This reminds us of the procedure described earlier in connection with deriving an echelon matrix from a given matrix. If we can find a sequence of row operations that transforms the given matrix $\mathbf{A}_{n \times n}$ into **I**, then the premultiplying matrix that represents this sequence must be the $\mathbf{B}_{n \times n}$ matrix we are looking for. Let us see whether the method works and, if so, how.

Suppose the given square matrix is

$$\mathbf{A} = \begin{bmatrix} 1 & 1 \\ 3 & 4 \end{bmatrix}$$

The first part of our procedure is to find a sequence of elementary row operations that transforms **A** into an echelon form. In the present case this task is easily accomplished: Subtract three times the first row from the second or, which is the same thing, premultiply **A** by

$$\mathbf{E}_1 = \begin{bmatrix} 1 & 0 \\ -3 & 1 \end{bmatrix}$$

This yields

$$\mathbf{E}_1\mathbf{A} = \begin{bmatrix} 1 & 0 \\ -3 & 1 \end{bmatrix} \begin{bmatrix} 1 & 1 \\ 3 & 4 \end{bmatrix} = \begin{bmatrix} 1 & 1 \\ 0 & 1 \end{bmatrix}$$

Note that this echelon matrix does not have any row consisting entirely of zeros; there are 1's all the way from top-left down to bottom-right of the principal diagonal; also the elementary row operator created zero elements everywhere below the principal diagonal. What remains now is to perform additional row operations so that all entries above the principal diagonal become zero. In the present case, this is accomplished by subtracting the second row from the first or, which is the same thing, premultiplying by

$$\mathbf{E}_2 = \begin{bmatrix} 1 & -1 \\ 0 & 1 \end{bmatrix}$$

giving

$$\mathbf{E}_2\mathbf{E}_1\mathbf{A} = \begin{bmatrix} 1 & -1 \\ 0 & 1 \end{bmatrix} \begin{bmatrix} 1 & 1 \\ 0 & 1 \end{bmatrix} = \begin{bmatrix} 1 & 0 \\ 0 & 1 \end{bmatrix}$$

The product

$$\mathbf{E} = \mathbf{E}_2\mathbf{E}_1 = \begin{bmatrix} 1 & -1 \\ 0 & 1 \end{bmatrix} \begin{bmatrix} 1 & 0 \\ -3 & 1 \end{bmatrix} = \begin{bmatrix} 4 & -1 \\ -3 & 1 \end{bmatrix}$$

(noting carefully the order in which they are entered—$\mathbf{E}_2$ on the left of $\mathbf{E}_1$) is the matrix **B** we are looking for, satisfying the relation $\mathbf{BA} = \mathbf{I}$. Let us check this out:

$$\begin{bmatrix} 4 & -1 \\ -3 & 1 \end{bmatrix} \begin{bmatrix} 1 & 1 \\ 3 & 4 \end{bmatrix} = \begin{bmatrix} 1 & 0 \\ 0 & 1 \end{bmatrix}$$

To consider another example, suppose we are given $\mathbf{G} = \begin{bmatrix} 5 & 7 \\ 2 & 3 \end{bmatrix}$. The sequence of operations that transforms **G** to an identity matrix is shown in Table 2.1.

$$\mathbf{E} = \mathbf{E}_4\mathbf{E}_3\mathbf{E}_2\mathbf{E}_1 = \begin{bmatrix} 3 & -7 \\ -2 & 5 \end{bmatrix}$$

Thus, in this case the matrix that satisfies the premultiplication requirement is

$$\begin{bmatrix} 3 & -7 \\ -2 & 5 \end{bmatrix}$$

We check this out by taking the product

$$\begin{bmatrix} 3 & -7 \\ -2 & 5 \end{bmatrix} \begin{bmatrix} 5 & 7 \\ 2 & 3 \end{bmatrix} = \begin{bmatrix} 1 & 0 \\ 0 & 1 \end{bmatrix}$$

TABLE 2.1

Description of Row Operation	*Elementary Row Operator*	*Resulting Matrix*
1. —	—	$\mathbf{G} = \begin{bmatrix} 5 & 7 \\ 2 & 3 \end{bmatrix}$
2. Multiply row 1 of **G** by 1/5	$\mathbf{E}_1 = \begin{bmatrix} 1/5 & 0 \\ 0 & 1 \end{bmatrix}$	$\mathbf{G}_1 = \begin{bmatrix} 1 & 7/5 \\ 2 & 3 \end{bmatrix}$
3. Subtract twice row 1 of $\mathbf{G}_1$ from its row 2	$\mathbf{E}_2 = \begin{bmatrix} 1 & 0 \\ -2 & 1 \end{bmatrix}$	$\mathbf{G}_2 = \begin{bmatrix} 1 & 7/5 \\ 0 & 1/5 \end{bmatrix}$
4. Multiply row 2 of $\mathbf{G}_2$ by 5	$\mathbf{E}_3 = \begin{bmatrix} 1 & 0 \\ 0 & 5 \end{bmatrix}$	$\mathbf{G}_3 = \begin{bmatrix} 1 & 7/5 \\ 0 & 1 \end{bmatrix}$
5. Subtract (7/5)th of row 2 of $\mathbf{G}_3$ from its row 1	$\mathbf{E}_4 = \begin{bmatrix} 1 & -7/5 \\ 0 & 1 \end{bmatrix}$	$\mathbf{G}_4 = \begin{bmatrix} 1 & 0 \\ 0 & 1 \end{bmatrix}$

It is easily verified that in both of the illustrative cases presented above, the matrix, which satisfies the premultiplication requirements, also satisfies the postmultiplication requirement:

$$\begin{bmatrix} 1 & 1 \\ 3 & 4 \end{bmatrix} \begin{bmatrix} 4 & -1 \\ -3 & 1 \end{bmatrix} = \begin{bmatrix} 1 & 0 \\ 0 & 1 \end{bmatrix}$$

and

$$\begin{bmatrix} 5 & 7 \\ 2 & 3 \end{bmatrix} \begin{bmatrix} 3 & -7 \\ -2 & 5 \end{bmatrix} = \begin{bmatrix} 1 & 0 \\ 0 & 1 \end{bmatrix}$$

In fact, it can be shown mathematically that this holds true in general. That is, given a square matrix **A**, if its inverse exists, the matrix **B**—constructed using elementary row operations on **A** so as to satisfy the premultiplication requirement **BA** = **I**—also satisfies the postmultiplication requirement **AB** = **I**. We shall not present the mathematical proof here.

There are a number of other ways to compute the inverse of a square matrix, if it exists. They are not described here. Computer programs are readily available for matrix inversion. Thus, for example, in the matrix procedure of SAS, the statements

A = –2 0 1 / –3 1 4 / –5 0 2;

B = INV (**A**);

set up the 3 × 3 matrix

$$\mathbf{A} = \begin{bmatrix} -2 & 0 & 1 \\ -3 & 1 & 4 \\ -5 & 0 & 2 \end{bmatrix},$$

compute its inverse, and print out the result as the matrix **B**. (The slashes in the first statement separate one row of **A** from the next.)

Although numerical methods are not emphasized here, it is important to know how to characterize square matrices that have inverses. We shall address this topic in the next chapter. In the remaining part of the present chapter we shall be concerned with illustrating some of the uses of the inverse of a square matrix. A conventional notation for the inverse of a square matrix **A** is $\mathbf{A}^{-1}$, which is read "**A** inverse" not "**A** to the –1 power."

It may also be noted, before turning to applications, that if a square matrix has an inverse, then this inverse is unique. That is, if **B** and **C** are two square matrices such that **AB** = **BA** = **I** and **AC** = **CA** = **I**, each matrix being of order $n \times n$, then **B** = **C**. To see that this is so, premultiply **AB** = **I** by **C**. This gives **CAB** = **CI**. But **CI** = **C** and we are told that **CA** = **I**. Hence, **CAB** = **CI** is equivalent to **IB** = **C** which is the same as **B** = **C**.

Application of the Inverse of a Matrix to the Solution of a System of Equations

We have already seen that a system of linear equations can be compactly expressed as a single matrix equation. Thus, the following two equations in two unknowns

$$2x + 3y = 1$$
$$4x + 9y = 2$$

can be written as

$$\begin{bmatrix} 2 & 3 \\ 4 & 9 \end{bmatrix} \begin{bmatrix} x \\ y \end{bmatrix} = \begin{bmatrix} 1 \\ 2 \end{bmatrix} \qquad [2.13]$$

If $\begin{bmatrix} 2 & 3 \\ 4 & 9 \end{bmatrix}^{-1}$ exists, premultiplication of both sides of 2.13 gives

$$\begin{bmatrix} 2 & 3 \\ 4 & 9 \end{bmatrix}^{-1} \begin{bmatrix} 2 & 3 \\ 4 & 9 \end{bmatrix} \begin{bmatrix} x \\ y \end{bmatrix} = \begin{bmatrix} 2 & 3 \\ 4 & 9 \end{bmatrix}^{-1} \begin{bmatrix} 1 \\ 2 \end{bmatrix}$$

that is

$$\begin{bmatrix} 1 & 0 \\ 0 & 1 \end{bmatrix} \begin{bmatrix} x \\ y \end{bmatrix} = \begin{bmatrix} 2 & 3 \\ 4 & 9 \end{bmatrix}^{-1} \begin{bmatrix} 1 \\ 2 \end{bmatrix}$$

or

$$\begin{bmatrix} x \\ y \end{bmatrix} = \begin{bmatrix} 2 & 3 \\ 4 & 9 \end{bmatrix}^{-1} \begin{bmatrix} 1 \\ 2 \end{bmatrix}$$

thus giving the solution required.

In the present case it is easy to verify that

$$\begin{bmatrix} 9/6 & -3/6 \\ -4/6 & 2/6 \end{bmatrix} \text{ is the inverse of } \begin{bmatrix} 2 & 3 \\ 4 & 9 \end{bmatrix}.$$

Hence, the required solution is

$$\begin{bmatrix} x \\ y \end{bmatrix} = \begin{bmatrix} 9/6 & -3/6 \\ -4/6 & 2/6 \end{bmatrix} \begin{bmatrix} 1 \\ 2 \end{bmatrix} = \begin{bmatrix} 1/2 \\ 0 \end{bmatrix}$$

or $x = ½; y = 0$. If we check this answer by direct substitution in the initial set of equations, we get

$$1 + 0 = 1$$
$$2 + 0 = 2$$

thus demonstrating that this is indeed a solution of the system.

As another example, consider the following system of three equations in three unknowns:

$$x + 2y + 3z = 1/2$$
$$x + 3y + 5z = 1$$
$$2x + 5y + 9z = 3/2$$

Writing this system as a matrix equation, we have

$$\begin{bmatrix} 1 & 2 & 3 \\ 1 & 3 & 5 \\ 2 & 5 & 9 \end{bmatrix} \begin{bmatrix} x \\ y \\ z \end{bmatrix} = \begin{bmatrix} 1/2 \\ 1 \\ 3/2 \end{bmatrix}$$

If we premultiply both sides of this equation by the inverse of

$$\begin{bmatrix} 1 & 2 & 3 \\ 1 & 3 & 5 \\ 2 & 5 & 9 \end{bmatrix}, \text{ namely, } \begin{bmatrix} 2 & -3 & 1 \\ 1 & 3 & -2 \\ -1 & -1 & 1 \end{bmatrix}$$

we get

$$\begin{bmatrix} 2 & -3 & 1 \\ 1 & 3 & -2 \\ -1 & -1 & 1 \end{bmatrix} \begin{bmatrix} 1 & 2 & 3 \\ 1 & 3 & 5 \\ 2 & 5 & 9 \end{bmatrix} \begin{bmatrix} x \\ y \\ z \end{bmatrix} = \begin{bmatrix} 2 & -3 & 1 \\ 1 & 3 & -2 \\ -1 & -1 & 1 \end{bmatrix} \begin{bmatrix} 1/2 \\ 1 \\ 3/2 \end{bmatrix}$$

which is

$$\begin{bmatrix} 1 & 0 & 0 \\ 0 & 1 & 0 \\ 0 & 0 & 1 \end{bmatrix} \begin{bmatrix} x \\ y \\ z \end{bmatrix} = \begin{bmatrix} -1/2 \\ 1/2 \\ 0 \end{bmatrix}$$

giving the solution $x = -1/2; y = 1/2; z = 0$, which on direct substitution in the original equations yields

$$\begin{aligned} -1/2 + 1 + 0 &= 1/2 \\ -1/2 + 3/2 + 0 &= 1 \\ -1 + 5/2 + 0 &= 3/2 \end{aligned}$$

thus proving that $x = -1/2; y = 1/2; z = 0$ is indeed a solution to the given system of equations.

Generalizing from these examples, if $\mathbf{A}$ is an $n \times n$ matrix, and $\mathbf{x}$ and $\mathbf{b}$ are both column vectors having n elements, the former consisting of unknowns and the latter of known constants, then a solution to the system of equations $\mathbf{Ax} = \mathbf{b}$ can be obtained by premultiplying both sides of the equation by $\mathbf{A}^{-1}$, if it exists. This is so because

$$\mathbf{A}^{-1}\mathbf{Ax} = \mathbf{A}^{-1}\mathbf{b}$$

is the same as

$$\mathbf{Ix} = \mathbf{A}^{-1}\mathbf{b},$$

by virtue of the definition of the inverse ($\mathbf{A}^{-1}\mathbf{A} = \mathbf{I}$), and this in turn is the same as

$$\mathbf{x} = \mathbf{A}^{-1}\mathbf{b}$$

by virtue of the definition of the identity matrix. That $\mathbf{x} = \mathbf{A}^{-1}\mathbf{b}$ satisifies the given equation system can be seen by substitution: $\mathbf{A}(\mathbf{A}^{-1}\mathbf{b}) = \mathbf{Ib} = \mathbf{b}$.

Application in Regression Analysis

Consider the following data on the height and weight of children:

Child number	1	2	3	4	5
Weight: Y	64	53	67	58	51
Height: X	57	50	61	52	45

Suppose we are interested in determining by the method of least squares the linear regression relationship giving weight in terms of height. The observational equations are

$$\begin{aligned} 64 &= b_0 + 57b_1 + e_1 \\ 53 &= b_0 + 50b_1 + e_2 \\ 67 &= b_0 + 61b_1 + e_3 \\ 58 &= b_0 + 52b_1 + e_4 \\ 51 &= b_0 + 45b_1 + e_5 \end{aligned}$$

or typically

$$y_i = b_0 + x_i b_1 + e_i$$

The least-squares method of determining the linear relationship giving y in terms of x involves determining b_0 and b_1 such that

$$\Sigma_i(y_i - b_0 - x_i b_1)^2$$

is the smallest, the summation being over the observed cases (the five children in the present instance). It can be shown that this minimization process results in a system of two linear equations in b_0 and b_1, called *normal equations*. These equations are as follows:

$$(n)b_0 + \left(\sum_{i=1}^{n} x_i\right) b_1 = \sum_{i=1}^{n} y_i$$

$$\left(\sum_{i=1}^{n} x_i\right) b_0 + \left(\sum_{i=1}^{n} x_i^2\right) b_1 = \sum_{i=1}^{n} x_i y_i$$

where n stands for the total number of cases observed (= 5, in the present instance). In matrix form these equations become

$$\begin{bmatrix} n & \Sigma x_i \\ \Sigma x_i & \Sigma x_i^2 \end{bmatrix} \begin{bmatrix} b_0 \\ b_1 \end{bmatrix} = \begin{bmatrix} \Sigma y_i \\ \Sigma x_i y_i \end{bmatrix} \qquad [2.14]$$

The 2×2 matrix on the left of 2.14 can be seen as the following product of the two matrices:

$$\begin{bmatrix} 1 & 1 & \ldots & 1 \\ x_1 & x_2 & \ldots & x_n \end{bmatrix} \begin{bmatrix} 1 & x_1 \\ 1 & x_2 \\ \cdot & \cdot \\ \cdot & \cdot \\ \cdot & \cdot \\ 1 & x_n \end{bmatrix}$$

Similarly, the matrix on the right of 2.14 can be seen as the following product of two matrices:

$$\begin{bmatrix} 1 & 1 & \ldots & 1 \\ x_1 & x_2 & \ldots & x_n \end{bmatrix} \begin{bmatrix} y_2 \\ \cdot \\ \cdot \\ \cdot \\ y_n \end{bmatrix}$$

We may therefore write the normal equations in a compact form as follows:

$$\mathbf{X'Xb} = \mathbf{X'y}$$

where

$$\mathbf{X} = \begin{bmatrix} 1 & x_1 \\ 1 & x_2 \\ \cdot & \cdot \\ \cdot & \cdot \\ \cdot & \cdot \\ 1 & x_n \end{bmatrix} \qquad \mathbf{b} = \begin{bmatrix} b_0 \\ b_1 \end{bmatrix} \qquad \mathbf{y} = \begin{bmatrix} y_2 \\ \cdot \\ \cdot \\ \cdot \\ y_n \end{bmatrix}$$

and $\mathbf{X}'$ is the tranpose of $\mathbf{X}$ (that is, the ith row of $\mathbf{X}'$ is the transpose of the ith column of $\mathbf{X}$ and vice versa).

For the data given above

$$n = 5$$

$$\sum_{i=1}^{n} x_i = 265$$

$$\sum_{i=1}^{n} x_i^2 = 14199$$

$$\sum_{i=1}^{n} y_i = 293$$

$$\sum_{i=1}^{n} x_i y_i = 15696$$

Hence the normal equations are

$$(5)b_0 + (265)b_1 = 293$$
$$(265)b_0 + (14199)b_1 = 15696$$

which in matrix form become

$$\begin{bmatrix} 5 & 265 \\ 265 & 14199 \end{bmatrix} \begin{bmatrix} b_0 \\ b_1 \end{bmatrix} = \begin{bmatrix} 293 \\ 15696 \end{bmatrix}$$

The solution is

$$\begin{bmatrix} b_0 \\ b_1 \end{bmatrix} = \begin{bmatrix} 5 & 265 \\ 265 & 14199 \end{bmatrix}^{-1} \begin{bmatrix} 293 \\ 15696 \end{bmatrix}$$

$$= \begin{bmatrix} \frac{14199}{770} & \frac{-265}{770} \\ \frac{-265}{770} & \frac{5}{770} \end{bmatrix} \begin{bmatrix} 293 \\ 15696 \end{bmatrix}$$

$$= \begin{bmatrix} 1.1260 \\ 1.0844 \end{bmatrix}$$

or $b_0 = 1.1260$ and $b_1 = 1.0844$.

We have thus seen that when fitting a straight-line regression model

$$y = b_0 + xb_1 + e$$

to a set of data consisting of n pairs of observations on the variables X and Y, we can characterize the problem under consideration in terms of

- **y,** the $n \times 1$ vector of observations on Y;
- **X,** the $n \times 2$ matrix of independent variables, including the constant term;
- **b,** the 2×1 vector of parameters; and
- **e,** the $n \times 1$ vector of random errors.

We have also seen that the normal equation can be written compactly as

$$\mathbf{X'Xb} = \mathbf{X'y}$$

where $\mathbf{X'}$ is the transpose of $\mathbf{X}$.

In the multiple regression context, the matrix **X** and the vector **b** reflect the presence of two or more regressors, in addition to the constant term, and the corresponding number of parameters (regression coefficients). Thus, with p regressors

$$\mathbf{X}_{n\times(p+1)} = \begin{bmatrix} 1 & x_{11} & \ldots & x_{1p} \\ \cdot & & & \cdot \\ \cdot & & & \cdot \\ \cdot & & & \cdot \\ 1 & x_{n1} & \ldots & x_{np} \end{bmatrix}$$

and

$$\mathbf{b}_{(p+1)\times 1} = \begin{bmatrix} b_0 \\ b_1 \\ \cdot \\ \cdot \\ \cdot \\ b_p \end{bmatrix}$$

The observational equations are

$$\mathbf{y}_{n\times 1} = \mathbf{X}_{n\times(p+1)}\mathbf{b}_{(p+1)\times 1} + \mathbf{e}_{n\times 1}$$

which summarizes in a single compact statement the n equations

$$y_i = b_0 + x_{i1}b_1 + \ldots + x_{ip}b_p + e_i$$

$i = 1,2, \ldots ,n$. This equivalence results from the matrix operations

$$\mathbf{Xb} + \mathbf{e} = \begin{bmatrix} 1 & x_{11} & \ldots & x_{1p} \\ \cdot & & & \cdot \\ \cdot & & & \cdot \\ \cdot & & & \cdot \\ 1 & x_{n1} & \ldots & x_{np} \end{bmatrix} \begin{bmatrix} b_0 \\ b_1 \\ \cdot \\ \cdot \\ \cdot \\ b_p \end{bmatrix} + \begin{bmatrix} e_1 \\ \cdot \\ \cdot \\ \cdot \\ e_n \end{bmatrix}$$

$$= \begin{bmatrix} b_0 + x_{11}b_1 + \ldots + x_{1p}b_p \\ \cdot \quad \cdot \quad \cdot \quad \cdot \quad \cdot \quad \cdot \quad \cdot \quad \cdot \\ b_0 + x_{n1}b_1 + \ldots + x_{np}b_p \end{bmatrix} + \begin{bmatrix} e_1 \\ \cdot \\ \cdot \\ \cdot \\ e_n \end{bmatrix}$$

$$= \begin{bmatrix} b_0 + x_{11}b_1 + \ldots + x_{1p}b_p + e_1 \\ \cdot \quad \cdot \quad \cdot \quad \cdot \quad \cdot \quad \cdot \quad \cdot \quad \cdot \quad \cdot \quad \cdot \\ b_0 + x_{n1}b_1 + \ldots + x_{np}b_p + e_n \end{bmatrix}$$

The normal equations can be compactly written as

$$\mathbf{X'Xb} = \mathbf{X'y}$$

where $\mathbf{X'}$ is the transpose of $\mathbf{X}$. The least-squares estimates are obtained as

$$\mathbf{b} = (\mathbf{X'X})^{-1}\mathbf{X'y}$$

if $(\mathbf{X'X})^{-1}$ exists.

Application in Input-Output Analysis

Input-output analysis was invented by Wassily Leontief more than 50 years ago. The basic tool of the analysis is a table that shows the inputs (purchases) and outputs (sales) of various industries in a system. A two-industry table is shown in Table 2.2.

On the left side of the table is a list of industries, one in each row. The figures in the rows represent sales (outputs) of the respective industries. Across the top of the table, the same set of industries are listed, one per column. The figures in a column under an industry show the purchases of that particular industry. There are two additional columns. The first one shows what is called the final demand, which includes purchases by households and by government, exports, and accumulation to inventory. The last column shows total output. Thus, in the table shown above,

TABLE 2.2

	Purchases (inputs) in Millions of $			
	Industry 1	*Industry 2*	*Final Demand*	*Total Output*
Sales (outputs) in millions of $				
Industry 1	560	1080	1160	2800
Industry 2	280	1440	1880	3600

reading across row 1, we notice that industry 1 had a total output of \$2800 million, of which intraindustry sales accounted for \$560 million, sales to industry 2 accounted for \$1080 million, and sales to households etc., accounted for the rest (\$1160 million). Column 1 shows that intraindustry purchases (inputs) accounted for \$560 million, and purchases from industry 2 accounted for \$280 million.

From this table, one calculates what are known as *technological coefficients* by dividing the inputs to each industry by the total output of the given industry. The technological coefficients calculated thus from the data shown above are

	Industry 1	Industry 2
Industry 1	$\frac{560}{2800} = 0.2$	$\frac{1080}{3600} = 0.3$
Industry 2	$\frac{280}{2800} = 0.1$	$\frac{1440}{3600} = 0.4$

These coefficients can be interpreted as follows: Each dollar worth of production in industry 1 requires \$0.2 worth intraindustry purchase and \$0.1 worth purchase from industry 2; similarly, each dollar worth of output from industry 2 requires \$0.3 worth purchase from industry 1 and \$0.4 worth intraindustry purchase.

The determination of new output levels required of all industries to meet a change in the final demand is one of the main problems in input-output analysis.

Let us write the technological coefficients calculated above in matrix form

$$\mathbf{A} = \begin{bmatrix} a_{11} & a_{12} \\ a_{21} & a_{22} \end{bmatrix} = \begin{bmatrix} 0.2 & 0.3 \\ 0.1 & 0.4 \end{bmatrix}$$

and the total output given above as

$$\mathbf{x} = \begin{bmatrix} 2800 \\ 3600 \end{bmatrix}$$

If the final demand is represented as

$$\mathbf{d} = \begin{bmatrix} 1160 \\ 1880 \end{bmatrix}$$

we immediately see that the following relationship holds between **A**, **x**, and **d**:

$$\mathbf{x} = \mathbf{Ax} + \mathbf{d}$$

This is simply a compact form of expressing the set of industry-specific relationships:

Total output = interindustry and intraindustry needs + final demand

The question is, what must **x** be, given **A** and **d**? The equation

$$\mathbf{x} = \mathbf{Ax} + \mathbf{d}$$

can be written as

$$\mathbf{Ix} = \mathbf{Ax} + \mathbf{d}$$

which is equivalent to

$$(\mathbf{I} - \mathbf{A})\mathbf{x} = \mathbf{d}$$

If $(\mathbf{I} - \mathbf{A})^{-1}$ exists, the required solution is readily obtained as

$$\mathbf{x} = (\mathbf{I} - \mathbf{A})^{-1}\mathbf{d}$$

Thus, if $\mathbf{A} = \begin{bmatrix} 0.2 & 0.3 \\ 0.1 & 0.4 \end{bmatrix}$ as calculated above, and **d** is given to be $\begin{bmatrix} 1160 \\ 1880 \end{bmatrix}$, then the total output should be

$$\mathbf{x} = (\mathbf{I} - \mathbf{A})^{-1}\mathbf{d} = \begin{bmatrix} (1 - 0.2) & (0 - 0.3) \\ (0 - 0.1) & (1 - 0.4) \end{bmatrix}^{-1} \mathbf{d}$$

$$= \begin{bmatrix} 0.8 & -0.3 \\ -0.1 & 0.6 \end{bmatrix}^{-1} \begin{bmatrix} 1160 \\ 1880 \end{bmatrix}$$

$$= \begin{bmatrix} 4/3 & 2/3 \\ 2/9 & 16/9 \end{bmatrix} \begin{bmatrix} 1160 \\ 1880 \end{bmatrix}$$

$$= \begin{bmatrix} 2800 \\ 3600 \end{bmatrix}$$

which is the same as those given in the original data. Given any expected level of final demand, one can forecast the required output, provided **A**, the matrix of technological coefficients, is known. We simply pre-multiply the final demand vector by the inverse of (**I** – **A**). Thus, if

$$\mathbf{d} = \begin{bmatrix} 900 \\ 2700 \end{bmatrix}$$

$$\mathbf{x} = \begin{bmatrix} 4/3 & 2/3 \\ 2/9 & 16/9 \end{bmatrix} \begin{bmatrix} 900 \\ 2700 \end{bmatrix} = \begin{bmatrix} 3000 \\ 5000 \end{bmatrix}$$

3. MORE ABOUT SIMULTANEOUS LINEAR EQUATIONS

This chapter introduces two important concepts, that of linear dependence among a collection of vectors, and that of the rank of a matrix. The latter concept is used in discussing equation systems with no solution, one unique solution and infinitely many solutions. The concept of generalized inverse is introduced in connection with linear equation systems with infinitely many solutions.

Linear Dependence Among a Set of Vectors

In Chapter 1 it was mentioned that a matrix with only one row or only one column is called a vector and that, consistent with this notion, we may think of a (column) vector with n elements as an n-tuple of numbers arranged as a column

$$\mathbf{a} = \begin{bmatrix} a_1 \\ \cdot \\ \cdot \\ \cdot \\ a_n \end{bmatrix}$$

Linear combination. A sum of scalar multiples of vectors in a collection of vectors, all containing the same number of elements, is called a *linear combination* of vectors in the collection. Thus, given

$$\mathbf{a} = \begin{bmatrix} 3 \\ 5 \end{bmatrix} \quad \text{and} \quad \mathbf{b} = \begin{bmatrix} 2 \\ 1 \end{bmatrix}$$

any sum such as

$$k_1\mathbf{a} + k_2\mathbf{b}$$

where k_1 and k_2 are any two scalars, not both zero, is a linear combination of **a** and **b**.

Linear combinations of what are known as unit vectors are worth special mention. An n-tuple is called a *unit vector* if all except one of its elements are zero and one element is unity. Thus

$$\begin{bmatrix} 1 \\ 0 \\ 0 \end{bmatrix} \quad \begin{bmatrix} 0 \\ 1 \\ 0 \end{bmatrix} \quad \begin{bmatrix} 0 \\ 0 \\ 1 \end{bmatrix}$$

are all 3-element unit vectors. It is easy to see that among n-tuples there are n unit vectors. Among 4-element vectors there are 4 unit vectors; among 5-element vectors, there are 5 unit vectors; and so on. Any n-tuple can be written as a linear combination of the corresponding set of unit vectors. Thus

$$\begin{bmatrix} 3 \\ 12 \\ -5 \end{bmatrix} = 3 \begin{bmatrix} 1 \\ 0 \\ 0 \end{bmatrix} + 12 \begin{bmatrix} 0 \\ 1 \\ 0 \end{bmatrix} - 5 \begin{bmatrix} 0 \\ 0 \\ 1 \end{bmatrix}$$

Linearly dependent set of vectors. A collection (set) of n-tuples is said to be *linearly dependent* if at least one vector in the set can be expressed as a linear combination of the remaining vectors. For example, the collection (set) of vectors

$$\begin{bmatrix} 1 \\ 3 \\ 5 \end{bmatrix} \quad \begin{bmatrix} 2 \\ 6 \\ 10 \end{bmatrix} \quad \begin{bmatrix} 8 \\ 4 \\ 2 \end{bmatrix}$$

is linearly dependent because one of them is twice another:

$$\begin{bmatrix} 2 \\ 6 \\ 10 \end{bmatrix} = 2 \begin{bmatrix} 1 \\ 3 \\ 5 \end{bmatrix}$$

Linear dependence is more commonly defined as follows. A collection of vectors $\mathbf{a}_1, \ldots, \mathbf{a}_n$ is said to be linearly dependent, if there exist numbers (scalars) $k_1, k_2, \ldots, k_n$, not all zero, such that $k_1\mathbf{a}_1 + \ldots + k_n\mathbf{a}_n$ equals a zero vector (one consisting entirely of zeros). The two definitions are equivalent.

Suppose there exist $k_1, \ldots, k_n$ such that $k_1\mathbf{a}_1 + k_2\mathbf{a}_2 + \ldots + k_n\mathbf{a}_n = \mathbf{0}$, where $\mathbf{0}$ denotes a zero vector. Subtracting $k_1\mathbf{a}_1$ from both sides we get

$$-k_1\mathbf{a}_1 = k_2\mathbf{a}_2 + \ldots + k_n\mathbf{a}_n$$

If $k_1 \neq 0$, we may multiply both sides by $(-1/k_1)$, and this gives

$$\mathbf{a}_1 = \left(\frac{-k_2}{k_1}\right)\mathbf{a}_2 + \ldots + \left(\frac{-k_n}{k_1}\right)\mathbf{a}_n$$

which expresses $\mathbf{a}_1$ as a linear combination of $\mathbf{a}_2, \ldots, \mathbf{a}_n$. We could have carried out these steps by subtracting any $k_i\mathbf{a}_i$ for which $k_i \neq 0$.

One way to check whether a given collection of vectors is linearly dependent is the following: Suppose we want to check whether $\begin{bmatrix} 2 \\ 4 \end{bmatrix}$ and $\begin{bmatrix} 1 \\ 3 \end{bmatrix}$ are linearly dependent. If they are, we know that there exist, by definition, numbers k_1 and k_2, not both zero, such that

$$k_1 \begin{bmatrix} 2 \\ 4 \end{bmatrix} + k_2 \begin{bmatrix} 1 \\ 3 \end{bmatrix} = \begin{bmatrix} 0 \\ 0 \end{bmatrix} \qquad [3.1]$$

But this vector equation is equivalent to the following two equations in two unknowns (k_1 and k_2):

$$\begin{aligned} 2k_1 + k_2 &= 0 \\ 4k_1 + 3k_2 &= 0 \end{aligned}$$

solving which we find $k_1 = k_2 = 0$.

Thus, in the present instance, there are no k_1 and k_2, not both zero, such that 3.1 is satisfied. We therefore declare that the two given vectors are not linearly dependent, or that they are *linearly independent*, which is the same thing. (A collection of vectors is either linearly dependent or linearly independent.)

TABLE 3.1

$$\begin{bmatrix} 1 & 5/3 \\ 2 & 1 \\ -3 & 2 \end{bmatrix} \quad \begin{bmatrix} 1 & 5/3 \\ 0 & -7/3 \\ 0 & 7 \end{bmatrix} \quad \begin{bmatrix} 1 & 5/3 \\ 0 & 1 \\ 0 & 7 \end{bmatrix} \quad \begin{bmatrix} 1 & 5/3 \\ 0 & 1 \\ 0 & 0 \end{bmatrix}$$

$$\underset{\sim}{A}_1 \qquad \underset{\sim}{A}_2 \qquad \underset{\sim}{A}_3 \qquad \underset{\sim}{A}_4$$

Another way of checking whether a given collection of vectors is linearly dependent is the following. Concatenate the given collection of vectors into a matrix; derive an echelon form from it; if the echelon form has one or more rows containing nothing but zeros, declare the collection as linearly dependent. To illustrate, suppose we want to determine whether the following vectors are linearly dependent:

$$\begin{bmatrix} 3 \\ 5 \end{bmatrix} \quad \begin{bmatrix} 2 \\ 1 \end{bmatrix} \quad \begin{bmatrix} -3 \\ 2 \end{bmatrix}$$

Stacking the transposes of these vectors we get the following matrix:

$$\mathbf{A} = \begin{bmatrix} 3 & 5 \\ 2 & 1 \\ -3 & 2 \end{bmatrix}$$

Table 3.1 shows the transformation of **A** into echelon form. Notice that row 1 of $\mathbf{A}_1$ equals (1/3) times row 1 of **A**; row 2 of $\mathbf{A}_2$ = row 2 of $\mathbf{A}_1$ – (2) times row 1 of $\mathbf{A}_1$; row 3 of $\mathbf{A}_2$ = row 3 of $\mathbf{A}_1$ + (3) times row 1 of $\mathbf{A}_1$; row 2 of $\mathbf{A}_3$ = (–3/7) times row 2 of $\mathbf{A}_2$; and row 3 of $\mathbf{A}_4$ = row 3 of $\mathbf{A}_3$ – (7) times row 2 of $\mathbf{A}_3$.

$\mathbf{A}_4$ is in echelon form. Its third row consists entirely of zeros. We therefore declare that the rows of the original matrix, **A**, are linearly dependent. To show that there exist numbers k_1, k_2, and k_3, not all zero, such that in **A**

$$(k_1)\text{(row 1)} + (k_2)\text{(row 2)} + (k_3)\text{(row 3)} = \mathbf{0},$$

all we need to do now is to express the third row of $\mathbf{A}_4$ in terms of the rows of **A**. To do this we retrace the steps that led to $\mathbf{A}_4$ from **A**.

$$\begin{aligned} \text{row 3 of } \mathbf{A}_4 &= \text{row 3 of } \mathbf{A}_3 - (7)(\text{row 2 of } \mathbf{A}_3) \\ &= \text{row 3 of } \mathbf{A}_2 - \left(-\frac{3}{7}\right)(7)(\text{row 2 of } \mathbf{A}_2) \end{aligned}$$

$$= [\text{row 3 of } \mathbf{A}_1 + (3)(\text{row 1 of } \mathbf{A}_1)]$$
$$+ (3)[\text{row 2 of } \mathbf{A}_1 - (2)(\text{row 1 of } \mathbf{A}_1)]$$
$$= [\text{row 3 of } \mathbf{A} + \text{row 1 of } \mathbf{A}]$$
$$+ [(3) \text{ row 2 of } \mathbf{A} - (2)(\text{row 1 of } \mathbf{A})]$$
$$= \text{row 3 of } \mathbf{A} + (3)(\text{row 2 of } \mathbf{A}) - \text{row 1 of } \mathbf{A}.$$

Thus

$$[0 \quad 0] = [-3 \quad 2] + (3)[2 \quad 1] - [3 \quad 5] \qquad [3.2]$$

from which it follows that any row of **A** can be written as a linear combination of the remaining rows: For example, by adding [3 5] to both sides of 3.2, we get

$$[3 \quad 5] = [-3 \quad 2] + (3)[2 \quad 1]$$

which expresses row 1 of **A** as a linear combination of rows 2 and 3. Similarly,

$$[-3 \quad 2] = [3 \quad 5] - (3)[2 \quad 1]$$

and

$$[2 \quad 1] = (1/3)[3 \quad 5] - (1/3)[-3 \quad 2].$$

The point we have illustrated is that when a given matrix is transformed into an echelon form by means of elementary row transformations, if we get one or more rows containing only zeros, then the rows of the original matrix satisfy a relationship of the form

$$\Sigma k_i \text{ row } i = \mathbf{0}$$

where not all k_i's are zero.

The Rank of a Matrix

When a matrix is transformed into an echelon matrix by elementary row operations, the number of nonzero rows in the resulting echelon matrix is known as the *rank* of the original matrix. It is possible to show that no matter what particular elementary row operations we use in deriving an echelon matrix from a given matrix, the *number* of nonzero rows in the echelon matrix will be the same. The usual notation for the rank of a matrix **A** is $r(\mathbf{A})$. The rank is zero only for a zero (null) matrix. All other matrices have positive (greater than zero) rank.

EXAMPLES

1. The rank of $\begin{bmatrix} 1 & 2 \\ 3 & 5 \end{bmatrix}$ is 2, as can be verified by deriving an echelon matrix from it:

$$\begin{bmatrix} 1 & 2 \\ 3 & 5 \end{bmatrix} \quad \begin{bmatrix} 1 & 2 \\ 0 & -1 \end{bmatrix} \quad \begin{bmatrix} 1 & 2 \\ 0 & 1 \end{bmatrix}$$

2. The rank of $\begin{bmatrix} 1 & 2 \\ 2 & 4 \end{bmatrix}$ is 1 because there is only one nonzero row in the echelon form:

$$\begin{bmatrix} 1 & 2 \\ 2 & 4 \end{bmatrix} \quad \begin{bmatrix} 1 & 2 \\ 0 & 0 \end{bmatrix}$$

3. The rank of $\begin{bmatrix} 1 & 2 & 3 & 4 \\ 2 & 3 & 4 & 5 \end{bmatrix}$ is 2:

$$\begin{bmatrix} 1 & 2 & 3 & 4 \\ 2 & 3 & 4 & 5 \end{bmatrix} \quad \begin{bmatrix} 1 & 2 & 3 & 4 \\ 0 & -1 & -2 & -3 \end{bmatrix} \quad \begin{bmatrix} 1 & 2 & 3 & 4 \\ 0 & 1 & 2 & 3 \end{bmatrix}$$

4. The rank of $\begin{bmatrix} 1 & 2 \\ 2 & 3 \\ 4 & 5 \\ 6 & 7 \end{bmatrix}$ is 2:

$$\begin{bmatrix} 1 & 2 \\ 2 & 3 \\ 4 & 5 \\ 6 & 7 \end{bmatrix} \quad \begin{bmatrix} 1 & 2 \\ 0 & -1 \\ 0 & -3 \\ 0 & -5 \end{bmatrix} \quad \begin{bmatrix} 1 & 2 \\ 0 & 1 \\ 0 & -3 \\ 0 & -5 \end{bmatrix} \quad \begin{bmatrix} 1 & 2 \\ 0 & 1 \\ 0 & 0 \\ 0 & 0 \end{bmatrix}$$

In these examples we have been looking at given matrices as concatenations of row vectors. Let us see whether we would get equivalent results if we were to treat a matrix as a concatenation of a number of column vectors. Thus, instead of treating the matrix $\begin{bmatrix} 1 & 2 \\ 3 & 5 \end{bmatrix}$ as two vectors $[1 \quad 2]$ and $[3 \quad 5]$ stacked up, suppose we were to treat it as a concatenation of $\begin{bmatrix} 1 \\ 3 \end{bmatrix}$ and $\begin{bmatrix} 2 \\ 5 \end{bmatrix}$.

Elementary column operations give the following results:

$$\begin{bmatrix} 1 & 2 \\ 3 & 5 \end{bmatrix} \quad \begin{bmatrix} 1 & 0 \\ 3 & -1 \end{bmatrix} \quad \begin{bmatrix} 1 & 0 \\ 3 & 1 \end{bmatrix}$$

$$\mathbf{A} \qquad \mathbf{B}_1 \qquad \mathbf{B}_2$$

The number of nonzero columns in $\mathbf{B}_2$ is 2, which is equal to the rank of **A** obtained earlier (Example 1) using elementary row operations. This example illustrates the point that the rank of a matrix is the same whether we treat the matrix as a concatenation of rows or as one of columns. To give another illustration, consider the matrix used in Example 3 above:

$$\begin{bmatrix} 1 & 2 & 3 & 4 \\ 2 & 3 & 4 & 5 \end{bmatrix}$$

Elementary column operations give the following results:

$$\begin{bmatrix} 1 & 2 & 3 & 4 \\ 2 & 3 & 4 & 5 \end{bmatrix} \begin{bmatrix} 1 & 0 & 0 & 0 \\ 2 & -1 & -2 & -3 \end{bmatrix} \begin{bmatrix} 1 & 0 & 0 & 0 \\ 2 & 1 & -2 & -3 \end{bmatrix} \begin{bmatrix} 1 & 0 & 0 & 0 \\ 2 & 1 & 0 & 0 \end{bmatrix}$$

The number of nonzero columns in the last matrix is the same as the rank obtained earlier by elementary row operations.

Referring to echelon matrices derived by means of elementary row operations as row-echelon forms (matrices), and those derived by elementary column operations as column-echelon forms (matrices), we may now define the rank of a matrix as the *number of nonzero rows in a row-echelon form of the given matrix or as the number of nonzero columns in a column-echelon form.* We immediately notice that the rank of an $n \times m$ matrix exceeds neither n nor m.

There are other equivalent definitions of the rank of a matrix, which will not be explored in this paper. Interested readers may refer, for example, to Ben Noble's *Applied Linear Algebra* (1969).

The rank of a square matrix determines whether the matrix has an inverse. A square matrix of order $n \times n$ is said to be of *full rank* if its rank is n. A square matrix of full rank has an inverse; such matrices are said to be *nonsingular*. Square matrices with less than full rank are said to be *singular*, and they are not invertible.

Simultaneous Linear Equations

Using the ranks of two matrices associated with simultaneous linear equations, it is possible to say whether the equations have no solution, one solution, or infinitely many solutions.

Consider the following two equations in two unknowns:

$$\begin{aligned} 3x + y &= 7 \\ x + y &= 3 \end{aligned} \qquad [3.3]$$

Let us try the substitution method for solving these equations. From the second equation we have $y = 3 - x$, and on substitution for y in the first we get $3x + (3 - x) = 7$, which leads to the solution $x = 2, y = 1$. That these values satisfy 3.3 is easily verified. Thus we have shown that a solution exists in this case. The question whether this is the only solution possible remains to be answered. Before addressing this question, let us consider the equations 3.4:

$$\begin{aligned} 3x + y &= 7 \\ 6x + 2y &= 10 \end{aligned} \qquad [3.4]$$

Casual inspection tells us that there is a problem here: While the left-hand side of the second equation is twice the left-hand side of the first, this relationship does not hold between the right-hand sides of the two equations. The two equations in the system are therefore inconsistent. No values can be found for x and y which satisfy both equations; any values that satisfy one will not satisfy the other.

Now consider the equations

$$\begin{aligned} 3x + y &= 7 \\ 6x + 2y &= 14 \end{aligned} \qquad [3.5]$$

Here the left-hand sides of the two equations have the same relation as the right-hand sides; i.e., the second equation is double the first. Thus, in effect, we have only one equation. But if there is only one equation involving two unknowns we can have an infinite number of solutions. Thus, in the present case it is easy to verify that the values

$$\begin{aligned} x &= \theta \\ y &= 7 - 3\theta \end{aligned}$$

satisfy the equations 3.5, whatever the value of θ.

Writing the equation systems in matrix form, we find that 3.3, 3.4, and 3.5, respectively, are equivalent to

$$\begin{bmatrix} 3 & 1 \\ 1 & 1 \end{bmatrix} \begin{bmatrix} x \\ y \end{bmatrix} = \begin{bmatrix} 7 \\ 3 \end{bmatrix} \qquad [3.6]$$

$$\begin{bmatrix} 3 & 1 \\ 6 & 2 \end{bmatrix} \begin{bmatrix} x \\ y \end{bmatrix} = \begin{bmatrix} 7 \\ 10 \end{bmatrix} \quad [3.7]$$

$$\begin{bmatrix} 3 & 1 \\ 6 & 2 \end{bmatrix} \begin{bmatrix} x \\ y \end{bmatrix} = \begin{bmatrix} 7 \\ 14 \end{bmatrix} \quad [3.8]$$

With each of these equation systems, we can associate two matrices, the *coefficient matrix* and the *augmented matrix*, the latter having the vector of constant terms as its last column, and the coefficient matrix for the rest. Thus the coefficient matrix and the augmented matrix associated with 3.6 are, respectively,

$$\begin{bmatrix} 3 & 1 \\ 1 & 1 \end{bmatrix} \text{ and } \begin{bmatrix} 3 & 1 & 7 \\ 1 & 1 & 3 \end{bmatrix}$$

The corresponding ones associated with 3.7 are

$$\begin{bmatrix} 3 & 1 \\ 6 & 2 \end{bmatrix} \text{ and } \begin{bmatrix} 3 & 1 & 7 \\ 6 & 2 & 10 \end{bmatrix}$$

And those associated with 3.8 are

$$\begin{bmatrix} 3 & 1 \\ 6 & 2 \end{bmatrix} \text{ and } \begin{bmatrix} 3 & 1 & 7 \\ 6 & 2 & 14 \end{bmatrix}$$

In the case of 3.6 we notice that both the coefficient matrix and the augmented matrix are of rank 2:

$$\begin{bmatrix} 3 & 1 \\ 1 & 1 \end{bmatrix} \quad \begin{bmatrix} 1 & 1/3 \\ 1 & 1 \end{bmatrix} \quad \begin{bmatrix} 1 & 1/3 \\ 0 & 2/3 \end{bmatrix} \quad \begin{bmatrix} 1 & 1/3 \\ 0 & 1 \end{bmatrix}$$

$$\begin{bmatrix} 3 & 1 & 7 \\ 1 & 1 & 3 \end{bmatrix} \quad \begin{bmatrix} 1 & 1/3 & 7/3 \\ 1 & 1 & 3 \end{bmatrix} \quad \begin{bmatrix} 1 & 1/3 & 7/3 \\ 0 & 2/3 & 2/3 \end{bmatrix} \quad \begin{bmatrix} 1 & 1/3 & 7/3 \\ 0 & 1 & 1 \end{bmatrix}$$

In the case of 3.7, however, the coefficient matrix is of rank 1, while the augmented matrix is of rank 2:

$$\begin{bmatrix} 3 & 1 \\ 6 & 2 \end{bmatrix} \quad \begin{bmatrix} 1 & 1/3 \\ 6 & 2 \end{bmatrix} \quad \begin{bmatrix} 1 & 1/3 \\ 0 & 0 \end{bmatrix}$$

$$\begin{bmatrix} 3 & 1 & 7 \\ 6 & 2 & 10 \end{bmatrix} \quad \begin{bmatrix} 1 & 1/3 & 7/3 \\ 6 & 2 & 10 \end{bmatrix} \quad \begin{bmatrix} 1 & 1/3 & 7/3 \\ 0 & 0 & -4 \end{bmatrix} \quad \begin{bmatrix} 1 & 1/3 & 7/3 \\ 0 & 0 & 1 \end{bmatrix}$$

And in the case of 3.7 both matrices are of rank 1:

$$\begin{bmatrix} 3 & 1 \\ 6 & 2 \end{bmatrix} \quad \begin{bmatrix} 1 & 1/3 \\ 6 & 2 \end{bmatrix} \quad \begin{bmatrix} 1 & 1/3 \\ 0 & 0 \end{bmatrix}$$

$$\begin{bmatrix} 3 & 1 & 7 \\ 6 & 2 & 14 \end{bmatrix} \quad \begin{bmatrix} 1 & 1/3 & 7/3 \\ 6 & 2 & 14 \end{bmatrix} \quad \begin{bmatrix} 1 & 1/3 & 7/3 \\ 0 & 0 & 0 \end{bmatrix}$$

Thus (1) the coefficient matrix may be of full rank or less than full rank, and (2) the rank of the coefficient matrix may be less than or equal to but never greater than that of the augmented matrix. Equation systems in which the coefficient matrix and the augmented matrix differ in their rank are said to be inconsistent. Such equation systems do not have any solution, and these systems will not be discussed further in this paper. An equation system that is consistent and whose coefficient matrix is of full rank has a solution, which can be shown to be unique. If the equation system is consistent, but the rank of the coefficient matrix is less than full, then the equation system has an infinite number of solutions.

THE FULL-RANK CASE

Since in the full-rank case the coefficient matrix is nonsingular, we can get a solution by premultiplying both sides of the equation in matrix version by the inverse of the coefficient matrix, as we have already seen in Chapter 2. That is, if the equation is $\mathbf{Ax} = \mathbf{b}$, a solution is $\mathbf{x} = \mathbf{A}^{-1}\mathbf{b}$. Thus, in the case of 3.3 a solution can be obtained as

$$\begin{bmatrix} x \\ y \end{bmatrix} = \begin{bmatrix} 3 & 1 \\ 1 & 1 \end{bmatrix}^{-1} \begin{bmatrix} 7 \\ 3 \end{bmatrix} = \begin{bmatrix} 1/2 & -1/2 \\ -1/2 & 3/2 \end{bmatrix} \begin{bmatrix} 7 \\ 3 \end{bmatrix} = \begin{bmatrix} 2 \\ 1 \end{bmatrix}$$

or $x = 2$; $y = 1$, which is the same as the solution obtained earlier by the method of substitution. That the solution of $\mathbf{Ax} = \mathbf{b}$ is unique if $\mathbf{A}^{-1}$ exists follows from the fact that if $\mathbf{x}_1$ and $\mathbf{x}_2$ are two solutions of $\mathbf{Ax} = \mathbf{b}$, then on substitution $\mathbf{Ax}_1 = \mathbf{b}$ and $\mathbf{Ax}_2 = \mathbf{b}$, which gives on subtraction $\mathbf{A}(\mathbf{x}_1 - \mathbf{x}_2) = \mathbf{0}$, a zero vector, which, in turn, gives, on premultiplication by $\mathbf{A}^{-1}$, $(\mathbf{x}_1 - \mathbf{x}_2) = \mathbf{0}$, implying that $\mathbf{x}_1 = \mathbf{x}_2$.

Before we discuss the case of less than full rank for the coefficient matrix, let us get acquainted with a solution procedure that can be adapted to both the full-rank case and the less than full-rank case.

The method consists of carrying out elementary row operations on the augmented matrix so as to obtain an identity matrix in the columns corresponding to the coefficient matrix. An illustration is shown in Table 3.2, using the equation system 3.3.

TABLE 3.2

Step 0	$\begin{bmatrix} 3 & 1 & 7 \\ 1 & 1 & 3 \end{bmatrix}$		
Step 1	$\begin{bmatrix} 1 & 1/3 & 7/3 \\ 1 & 1 & 3 \end{bmatrix}$		
Step 2	$\begin{bmatrix} 1 & 1/3 & 7/3 \\ 0 & 2/3 & 2/3 \end{bmatrix}$		
Step 3	$\begin{bmatrix} 1 & 1/3 & 7/3 \\ 0 & 1 & 1 \end{bmatrix}$		
Step 4	$\begin{bmatrix} 1 & 0 & 2 \\ 0 & 1 & 1 \end{bmatrix}$		

The solution appears in the final column in step 4. This method combines the two steps, finding $\mathbf{A}^{-1}$ and then obtaining the product $\mathbf{A}^{-1}\mathbf{b}$. Note that in preparing Table 3.2 at each step we get a transformed matrix from the one in the preceding step, by means of the familiar row operations. For example, the matrix in step 1 is obtained from the original matrix by multiplying the first row by (1/3); the one in step 2 is obtained by subtracting the first row from the second row of the one in step 1; and so on; finally, row 1 in step 4 is obtained by subtracting (1/3) times row 2 in step 3 from row 1 in step 3. In step 4 we notice that the columns corresponding to the original coefficient matrix form an identity matrix. At this step the final column gives the required solution.

THE LESS THAN FULL-RANK CASE AND THE GENERALIZED INVERSE

Let us now apply the same method to the equation system 3.5. (See Table 3.3.) We cannot go beyond step 2 in the table. The transformed matrix obtained in this step tells us that the original system of equations is equivalent to a single equation:

$$x + (1/3)y = 7/3$$

TABLE 3.3

Step 0	$\begin{bmatrix} 3 & 1 & 7 \\ 6 & 2 & 14 \end{bmatrix}$
Step 1	$\begin{bmatrix} 1 & 1/3 & 7/3 \\ 6 & 2 & 14 \end{bmatrix}$
Step 2	$\begin{bmatrix} 1 & 1/3 & 7/3 \\ 0 & 0 & 0 \end{bmatrix}$

which can be written as

$$x = (7/3) - (1/3)y$$

In effect we have solved the equation system. We notice that if we let y take any value we can then determine x accordingly. We may therefore treat y as a parameter, say ϕ where ϕ is any number—positive, negative, or zero. In terms of ϕ, the solution to the original equation system is

$$\begin{aligned} x &= (7/3) - (1/3)\phi \\ y &= \phi \end{aligned} \qquad [3.9]$$

For example, one solution occurs if $\phi = 0$, which gives $x = 7/3$, $y = 0$; another, if $\phi = 1$, which yields $x = 2$, $y = 1$; and so on. More generally, we can convince ourselves that $x = (7/3) - (1/3)\phi$, $y = \phi$ is a solution whatever be ϕ, by substituting these values in the original equation:

$$\begin{aligned} 3[(7/3) - (1/3)\phi] + \phi &= 7 \\ 6[(7/3) - (1/3)\phi] + 2\phi &= 14 \end{aligned}$$

The point to note is that in this case the solution is not unique and that, since ϕ may take any value, we have an infinite number of solutions.

Earlier we obtained by the method of substitution the following solution to the equation system 3.5:

$$\begin{aligned} x &= \theta \\ y &= 7 - 3\theta \end{aligned} \qquad [3.10]$$

TABLE 3.4

$$\text{Step 0} \quad \begin{bmatrix} 1 & 1 & 1 & 5 \\ 1 & 2 & -1 & 0 \\ 2 & 3 & 0 & 5 \end{bmatrix}$$

$$\text{Step 1} \quad \begin{bmatrix} 1 & 1 & 1 & 5 \\ 0 & 1 & -2 & -5 \\ 2 & 3 & 0 & 5 \end{bmatrix}$$

$$\text{Step 2} \quad \begin{bmatrix} 1 & 1 & 1 & 5 \\ 0 & 1 & -2 & -5 \\ 0 & 1 & -2 & -5 \end{bmatrix}$$

$$\text{Step 3} \quad \begin{bmatrix} 1 & 1 & 1 & 5 \\ 0 & 1 & -2 & -5 \\ 0 & 0 & 0 & 0 \end{bmatrix}$$

$$\text{Step 4} \quad \begin{bmatrix} 1 & 0 & 3 & 10 \\ 0 & 1 & -2 & -5 \\ 0 & 0 & 0 & 0 \end{bmatrix}$$

At first glance it may appear that the solution 3.9 is different from 3.10. But in fact the two are equivalent, since any value of θ corresponds to a single value of ϕ and vice versa. If we write $\phi = 7 - 3\theta$, 3.10 becomes the same as 3.9. To give a numerical example, suppose we give ϕ the value 0, then $\theta = 7/3$ and, according to 3.9, the solution is $x = (7/3), y = 0$; which is the same as the solution according to 3.10.

Let us consider another example involving the less than full-rank case. Suppose we want to solve the equation system

$$\begin{aligned} x + y + z &= 5 \\ x + 2y - z &= 0 \\ 2x + 3y &= 5 \end{aligned} \qquad [3.11]$$

This is a less than full-rank case because the third equation is the sum of the first two. Carry out the computations precisely in the manner as before. Remember that our objective is to transform the columns corresponding to the coefficient matrix into those of an identity matrix *as far as possible*. Table 3.4 shows the computations. The matrix in step 1 is obtained from the original augmented matrix (shown in step 0) by subtracting row 1 from row 2. Row 3 of the matrix in step 1 minus (2) times its row 1 gives row 3 of the matrix in step 2. Row 3 of the matrix in step 2 minus its row 2 gives row 3 of the matrix in step 3. And row 1 of the matrix in step 3 minus its row 2 gives row 1 of the matrix in step 4.

Step 4 is as far as we can go. The matrix in this step tells us that the original equation system is equivalent to the following:

$$\begin{bmatrix} 1 & 0 & 3 \\ 0 & 1 & -2 \\ 0 & 0 & 0 \end{bmatrix} \begin{bmatrix} x \\ y \\ z \end{bmatrix} = \begin{bmatrix} 10 \\ -5 \\ 0 \end{bmatrix}$$

or

$$\begin{aligned} x + 3z &= 10 \\ y - 2z &= -5 \end{aligned}$$

Assigning parametric value $z = \theta$, we get the general solution

$$\begin{bmatrix} x \\ y \\ z \end{bmatrix} = \begin{bmatrix} 10 - 3\theta \\ -5 + 2\theta \\ \theta \end{bmatrix} \qquad [3.12]$$

(Check for yourself that this solution satisfies 3.11: $(10 - 3\theta) + (-5 + 2\theta) + \theta = 5$, whatever be θ. $(10 - 3\theta) + 2(-5 + 2\theta) - \theta = 0$, for all values of θ. $2(10 - 3\theta) + 3(-5 + 2\theta) = 5$, irrespective of the value of θ.)

In these computations we did not explicitly make use of the fact that the third equation in the initial set is the sum of the first two. Suppose we decide to use the fact right from the start. Thus, suppose we decide to ignore the third equation on the ground that it is redundant, given the first two. This means we are left with just two equations (in three unknowns):

$$\begin{aligned} x + y + z &= 5 \\ x + 2y - z &= 0 \end{aligned} \qquad [3.13]$$

Let us carry out the computations as before. Referring to Table 3.5, the matrix in step 2 tells us that the original equation system is equivalent to

$$\begin{bmatrix} 1 & 0 & 3 \\ 0 & 1 & -2 \end{bmatrix} \begin{bmatrix} x \\ y \\ z \end{bmatrix} = \begin{bmatrix} 10 \\ -5 \end{bmatrix} \qquad [3.14]$$

or

$$\begin{aligned} x + 3z &= 10 \\ y - 2z &= -5 \end{aligned} \qquad [3.15]$$

TABLE 3.5

Step 0	1	1	1	5
	1	2	−1	0
Step 1	1	1	1	5
	0	1	−2	−5
Step 2	1	0	3	10
	0	1	−2	−5

Parameterizing z (= θ) now gives the general solution

$$\begin{bmatrix} x \\ y \\ z \end{bmatrix} = \begin{bmatrix} 10 - 3\theta \\ -5 + 2\theta \\ \theta \end{bmatrix} \qquad [3.16]$$

Verify for yourself that you would have obtained these same general solutions if you had started with any other pair of equations from 3.11. Also of interest is to write a general solution, treating y (= ϕ) as a parameter. From 3.15 show that if $y = \phi$, $z = (5 + \phi)/2$ and $x = (5 - 3\phi)/2$, and verify that these values satisfy the original equations, whatever be ϕ.

Another interesting point is that if we had parameterized z to begin with, equation system 3.3 would have been

$$\begin{aligned} x + y &= 5 - \theta \\ x + 2y &= \theta \end{aligned}$$

and instead of Table 3.5 we would have had Table 3.5a, leading straight to the solution

$$x = 10 - 3\theta, \qquad y = -5 + 2\theta, \qquad z = \theta$$

which is the same as 3.16.

This latter computation suggests the following strategy for solving simultaneous linear equations in the less than full-rank case.

(1) Delete redundant equations. If there are n equations in n unknowns, and the rank of the coefficient matrix $\mathbf{A}$ is $r(\mathbf{A})$ then there are $n - r(\mathbf{A})$ redundant equations.

TABLE 3.5a

Step 0	$\begin{bmatrix} 1 & 1 & (5-\theta) \\ 1 & 2 & \theta \end{bmatrix}$
Step 1	$\begin{bmatrix} 1 & 1 & (5-\theta) \\ 0 & 1 & (-5+2\theta) \end{bmatrix}$
Step 2	$\begin{bmatrix} 1 & 0 & (10-3\theta) \\ 0 & 1 & (-5+2\theta) \end{bmatrix}$

(2) Parameterize $n - r(\mathbf{A})$ unknowns, and rewrite the retained, $r(\mathbf{A})$, equations as equations in $r(\mathbf{A})$ unknowns, and solve by the method of solving the full-rank case.

Let us apply this latter technique to another equation system. Suppose we want to solve

$$\begin{aligned} x - y + 3z &= 2 \\ x + y + 2z &= 4 \\ 3x - y + 8z &= 8 \end{aligned} \qquad [3.17]$$

First we want to determine whether this is a less than full-rank case. The coefficient matrix is

$$\begin{bmatrix} 1 & -1 & 3 \\ 1 & 1 & 2 \\ 3 & -1 & 8 \end{bmatrix}$$

Subtracting the first row from the second and three times the first row from the third, we get

$$\begin{bmatrix} 1 & -1 & 3 \\ 0 & 2 & -1 \\ 0 & 2 & -1 \end{bmatrix}$$

which leads to the echelon form

$$\begin{bmatrix} 1 & -1 & 0 \\ 0 & 1 & -1/2 \\ 0 & 0 & 0 \end{bmatrix}$$

revealing that the rank of the coefficient matrix is 2. We have thus a less than full-rank case here.

Let us treat the third equation in 3.17 as redundant, and parameterize $z(=\theta)$. This yields the following equation system:

$$\begin{bmatrix} 1 & -1 \\ 1 & 1 \end{bmatrix} \begin{bmatrix} x \\ y \end{bmatrix} = \begin{bmatrix} 2-3\theta \\ 4-2\theta \end{bmatrix} \qquad [3.18]$$

whose solution is

$$\begin{bmatrix} x \\ y \end{bmatrix} = \begin{bmatrix} 1 & -1 \\ 1 & 1 \end{bmatrix}^{-1} \begin{bmatrix} 2-3\theta \\ 4-2\theta \end{bmatrix}$$

$$= \begin{bmatrix} 1/2 & 1/2 \\ -1/2 & 1/2 \end{bmatrix} \begin{bmatrix} 2-3\theta \\ 4-2\theta \end{bmatrix}$$

$$= \begin{bmatrix} 3-(5/2)\theta \\ 1+(1/2)\theta \end{bmatrix} \qquad [3.19]$$

so that the general solution is

$$\begin{bmatrix} x \\ y \\ z \end{bmatrix} = \begin{bmatrix} 3-(5/2)\,\theta \\ 1+(1/2)\,\theta \\ \theta \end{bmatrix} \qquad [3.20]$$

If we set $\theta = 0$ in 3.20, we get a particular solution

$$\begin{bmatrix} x \\ y \\ z \end{bmatrix} = \begin{bmatrix} 3 \\ 1 \\ 0 \end{bmatrix} \qquad [3.21]$$

which can be expressed in the form

$$\mathbf{x} = \mathbf{Gb} \qquad [3.22]$$

where

$$\mathbf{x} = \begin{bmatrix} x \\ y \\ z \end{bmatrix} \quad \mathbf{G} = \begin{bmatrix} \begin{bmatrix} 1 & -1 \\ 1 & 1 \end{bmatrix}^{-1} & \begin{matrix} 0 \\ 0 \end{matrix} \\ \begin{matrix} 0 & 0 \end{matrix} & 0 \end{bmatrix} = \begin{bmatrix} ½ & ½ & 0 \\ -½ & ½ & 0 \\ 0 & 0 & 0 \end{bmatrix} \quad \text{and} \quad \mathbf{b} = \begin{bmatrix} 2 \\ 4 \\ 8 \end{bmatrix}$$

Notice the connection between **G** and

$$\mathbf{A} = \begin{bmatrix} 1 & -1 & 3 \\ 1 & 1 & 2 \\ 3 & -1 & 8 \end{bmatrix}$$

the coefficient matrix of the original system of equations 3.17. (To get **G** from **A** we apply a three-step procedure. First we delete the third row and third column of **A**, then invert the resulting matrix, and finally border the resulting inverse with a row and a column of zeros corresponding to the row and column deleted from **A** in the first step.)

We may think of **G** as a matrix derived from the coefficient matrix **A**, such that $\mathbf{x} = \mathbf{Gb}$ is a solution of $\mathbf{Ax} = \mathbf{b}$. We call **G** which has this property (namely that $\mathbf{x} = \mathbf{Gb}$ is a solution to $\mathbf{Ax} = \mathbf{b}$) a *generalized inverse* of **A**. The label "generalized inverse" is apt because it is applicable to the full-rank case as well as the less than full-rank case. In the former case, of course, the three-step procedure described above becomes a single-step procedure, because the original matrix is invertible.

At first the concept of generalized inverse may appear difficult to grasp, but familiarity with it is quite useful. With this in mind a few more examples are given below.

Suppose we want to solve the following equation system:

$$\begin{bmatrix} 3 & 2 & 1 \\ 2 & 3 & -1 \\ 1 & -1 & 2 \end{bmatrix} \begin{bmatrix} x \\ y \\ z \end{bmatrix} = \begin{bmatrix} 5 \\ 6 \\ -1 \end{bmatrix} \qquad [3.23]$$

It is not difficult to see that the rank of the coefficient matrix is 2. (The third equation is the difference between the other two.) Let us discard the third equation and set $z = 0$ in the remaining equations. This gives

$$\begin{bmatrix} 3 & 2 \\ 2 & 3 \end{bmatrix} \begin{bmatrix} x \\ y \end{bmatrix} = \begin{bmatrix} 5 \\ 6 \end{bmatrix} \qquad [3.24]$$

which is a full-rank case. The solution of 3.24 is

$$\begin{bmatrix} x \\ y \end{bmatrix} = \begin{bmatrix} 3 & 2 \\ 2 & 3 \end{bmatrix}^{-1} \begin{bmatrix} 5 \\ 6 \end{bmatrix} = \begin{bmatrix} 0.6 & -0.4 \\ -0.4 & 0.6 \end{bmatrix} \begin{bmatrix} 5 \\ 6 \end{bmatrix} \qquad [3.25]$$

which can be written as shown below:

$$\begin{bmatrix} x \\ y \\ z \end{bmatrix} = \begin{bmatrix} 0.6 & -0.4 & 0.0 \\ -0.4 & 0.6 & 0.0 \\ 0.0 & 0.0 & 0.0 \end{bmatrix} \begin{bmatrix} 5 \\ 6 \\ -1 \end{bmatrix} \qquad [3.26]$$

giving $x = 0.6$, $y = 1.6$, and $z = 0$ as a solution of 3.23. (Verify that this indeed is a solution to the equation system.) In this case the 3×3 matrix on the right-hand side of 3.26 is a generalized inverse of the coefficient matrix of 3.23.

In the foregoing exercise we deleted the third equation in the system 3.23 and set $z = 0$. If instead we delete the first equation and set $x = 0$, we get from 3.23

$$\begin{bmatrix} 3 & -1 \\ -1 & 2 \end{bmatrix} \begin{bmatrix} y \\ z \end{bmatrix} = \begin{bmatrix} 6 \\ -1 \end{bmatrix}$$

which is a full-rank case. Its solution is

$$\begin{bmatrix} y \\ z \end{bmatrix} = \begin{bmatrix} 3 & -1 \\ -1 & 2 \end{bmatrix}^{-1} \begin{bmatrix} 6 \\ -1 \end{bmatrix}$$

$$= \begin{bmatrix} 0.4 & 0.2 \\ 0.2 & 0.6 \end{bmatrix} \begin{bmatrix} 6 \\ -1 \end{bmatrix} = \begin{bmatrix} 2.2 \\ 0.6 \end{bmatrix}$$

which is the same as

$$\begin{bmatrix} x \\ y \\ z \end{bmatrix} = \begin{bmatrix} 0.0 & 0.0 & 0.0 \\ 0.0 & 0.4 & 0.2 \\ 0.0 & 0.2 & 0.6 \end{bmatrix} \begin{bmatrix} 5 \\ 6 \\ -1 \end{bmatrix} = \begin{bmatrix} 0.0 \\ 2.2 \\ 0.6 \end{bmatrix} \qquad [3.27]$$

It is readily verified that the solution 3.27 does satisfy the original equation. According to our definition, the 3×3 matrix in 3.27 also qualifies to be called a generalized inverse of the coefficient matrix of 3.23. This illustrates the point that a given coefficient matrix can have more than one generalized inverse, of the kind defined above.

As another example, let us consider the system of equations

$$\begin{aligned} x + y + z &= 2 \\ x + y + 2z &= 1 \\ x + y + 3z &= 0 \end{aligned} \qquad [3.28]$$

The coefficient matrix is of rank 2. So we retain two equations and set one of the unknowns to zero. Let us retain the first and second equations. We may set equal to zero x or y, but not z. (If we set z equal to zero, the two-equation system thus obtained will not be of full rank. The decision as to which unknown is to be set equal to zero should be guided by this consideration. The objective is to derive a two-equation system in two unknowns with a unique solution.) Let us set $y = 0$. This gives

$$\begin{aligned} x + z &= 2 \\ x + 2z &= 1 \end{aligned}$$

the solution of which is

$$\begin{bmatrix} x \\ z \end{bmatrix} = \begin{bmatrix} 1 & 1 \\ 1 & 2 \end{bmatrix}^{-1} \begin{bmatrix} 2 \\ 1 \end{bmatrix}$$

$$= \begin{bmatrix} 2 & -1 \\ -1 & 1 \end{bmatrix} \begin{bmatrix} 2 \\ 1 \end{bmatrix} = \begin{bmatrix} 3 \\ -1 \end{bmatrix}$$

This is the same as

$$\begin{bmatrix} x \\ z \\ y \end{bmatrix} = \begin{bmatrix} 2 & -1 & 0 \\ -1 & 1 & 0 \\ 0 & 0 & 0 \end{bmatrix} \begin{bmatrix} 2 \\ 1 \\ 0 \end{bmatrix} \qquad [3.29]$$

which can be rearranged as

$$\begin{bmatrix} x \\ y \\ z \end{bmatrix} = \begin{bmatrix} 2 & -1 & 0 \\ 0 & 0 & 0 \\ -1 & 1 & 0 \end{bmatrix} \begin{bmatrix} 2 \\ 1 \\ 0 \end{bmatrix}$$

This rearrangement can be obtained by premultiplying both sides of 3.29 by the elementary matrix

$$\mathbf{E} = \begin{bmatrix} 1 & 0 & 0 \\ 0 & 0 & 1 \\ 0 & 1 & 0 \end{bmatrix}$$

In this case, obviously, according to our definition

$$\begin{bmatrix} 2 & -1 & 0 \\ 0 & 0 & 0 \\ -1 & 1 & 0 \end{bmatrix}$$

is a generalized inverse of

$$\begin{bmatrix} 1 & 1 & 1 \\ 1 & 1 & 2 \\ 1 & 1 & 3 \end{bmatrix}$$

The application of the generalized-inverse approach to the solution of a system of linear equations thus involves the following steps: Suppose the given equation system is $\mathbf{Ax} = \mathbf{b}$. Let $\mathbf{A}$ be of order $n \times n$, and let its rank be $r(\mathbf{A})$ or, more simply, r. The rank of the coefficient matrix is not affected by any rearrangement of the equations. The same is true of any alterations of the order in which the unknowns are written in the equations. If necessary, we can rearrange the equations and/or the order in which the unknowns appear in them, so that after the rearrangements we have an equation system, say, $\mathbf{A}_1\mathbf{x}_1 = \mathbf{b}_1$, whose coefficient matrix has its top-left r by r submatrix invertible. Let this submatrix be $\mathbf{B}$, and let $\mathbf{B}^{-1}$ be its inverse. We now border $\mathbf{B}^{-1}$ with $n-r$ columns of zeros on the right and $n-r$ rows of zeros at the bottom so as to produce a matrix of order $n \times n$. The matrix thus produced is a generalized inverse of $\mathbf{A}_1$ and by postmultiplying it by $\mathbf{b}_1$ we get a solution to $\mathbf{A}_1\mathbf{x}_1 = \mathbf{b}_1$. Premultiplying the generalized inverse of $\mathbf{A}_1$ obtained above by the elementary matrix $\mathbf{E}$, which is such that $\mathbf{Ex}_1 = \mathbf{x}$, we get a generalized inverse of $\mathbf{A}$.

A usual notation for a generalized inverse of a given matrix $\mathbf{A}$ is $\mathbf{A}^-$. The general solution to an equation system $\mathbf{Ax} = \mathbf{b}$ in terms of a generalized inverse of $\mathbf{A}$ is

$$\mathbf{x} = \mathbf{A}^- \mathbf{b} + (\mathbf{I} - \mathbf{A}^- \mathbf{A})\boldsymbol{\theta} \qquad [3.30]$$

where $\boldsymbol{\theta}$ is an arbitrary column vector.

To apply the formula 3.30 to, say, the equations system 3.17, we have from earlier calculations

$$\mathbf{A}^- = \begin{bmatrix} \frac{1}{2} & \frac{1}{2} & 0 \\ -\frac{1}{2} & \frac{1}{2} & 0 \\ 0 & 0 & 0 \end{bmatrix}$$

and, consequently,

$$\mathbf{A}^{-}\mathbf{b} = \begin{bmatrix} \tfrac{1}{2} & \tfrac{1}{2} & 0 \\ -\tfrac{1}{2} & \tfrac{1}{2} & 0 \\ 0 & 0 & 0 \end{bmatrix} \begin{bmatrix} 2 \\ 4 \\ 8 \end{bmatrix} = \begin{bmatrix} 3 \\ 1 \\ 0 \end{bmatrix}$$

and

$$\mathbf{A}^{-}\mathbf{A} = \begin{bmatrix} \tfrac{1}{2} & \tfrac{1}{2} & 0 \\ -\tfrac{1}{2} & \tfrac{1}{2} & 0 \\ 0 & 0 & 0 \end{bmatrix} \begin{bmatrix} 1 & -1 & 3 \\ 1 & 1 & 2 \\ 3 & -1 & 8 \end{bmatrix} = \begin{bmatrix} 1 & 0 & 5/2 \\ 0 & 1 & -1/2 \\ 0 & 0 & 0 \end{bmatrix}$$

so that

$$\mathbf{A}^{-}\mathbf{b} + (\mathbf{I} - \mathbf{A}^{-}\mathbf{A})\boldsymbol{\theta} = \begin{bmatrix} 3 - (5/2)\,\theta \\ 1 + (1/2)\,\theta \\ \theta \end{bmatrix} \qquad [3.31]$$

where

$$\boldsymbol{\theta} = \begin{bmatrix} \delta \\ \phi \\ \theta \end{bmatrix}$$

Notice that the solution 3.31 is the same as the one derived in 3.20.

The most common application of the generalized inverse in statistical analysis is in connection with linear models. Thus, in regression analyses involving categorical variables as regressors that are represented by dummy variables, the coefficient matrix of the normal equations has less than full rank. In the discussion of the solution to the normal equations in such circumstances, the concept of generalized inverse is invoked.

Homogeneous Equations

We now turn to a brief discussion of what are known as homogeneous linear equations. The equation system $\mathbf{Ax} = \mathbf{0}$, where $\mathbf{A}$ is the familiar coefficient matrix, $\mathbf{x}$ a vector of unknowns, and $\mathbf{0}$ a vector of zeros is a *homogeneous* system. Its name derives from the property that all nonzero terms in the system are similar in that each involves an unknown (variable) raised to the power one. Clearly, $\mathbf{x} = \mathbf{0}$ (i.e., a zero value for each unknown) satisfies the system of equations. Thus, given

$$x + y = 0$$
$$2x - y = 0$$

it is immediately clear that $x = y = 0$ is a solution to the system. Such a solution is usually called the *trivial solution*. An interesting question is whether there are circumstances under which a system of homogeneous equations has a solution other than the trivial one.

Consider the system of equations

$$\begin{aligned} x + y &= 0 \\ 2x - y &= 0 \\ 4x + y &= 0 \end{aligned} \qquad [3.32]$$

The coefficient matrix is

$$\begin{bmatrix} 1 & 1 \\ 2 & -1 \\ 4 & 1 \end{bmatrix}$$

an echelon form of which is

$$\begin{bmatrix} 1 & 1 \\ 0 & 1 \\ 0 & 0 \end{bmatrix}$$

indicating that the rank of the coefficient matrix is 2. The echelon form also tells us that the original equation system is equivalent to

$$\begin{aligned} x + y &= 0 \\ y &= 0 \end{aligned} \qquad [3.33]$$

leading to the only solution $x = y = 0$. It can be shown in general that a homogeneous system of linear equations $\mathbf{Ax} = \mathbf{0}$ involving n unknowns (variables), has one and only one solution—the trivial one, $\mathbf{x} = \mathbf{0}$—if the rank of the coefficient matrix is equal to the number of unknowns involved.

Now consider the following system of equations

$$\begin{aligned} x + y + z &= 0 \\ x - y - 2z &= 0 \\ 3x - y - 3z &= 0 \end{aligned}$$

The coefficient matrix is

$$\begin{bmatrix} 1 & 1 & 1 \\ 1 & -1 & -2 \\ 3 & -1 & -3 \end{bmatrix}$$

an echelon form of which is

$$\begin{bmatrix} 1 & 1 & 1 \\ 0 & 1 & 1.5 \\ 0 & 0 & 0 \end{bmatrix}$$

indicating that the coefficient matrix in the present instance is of rank 2. The echelon form also tells us that the original equation system is equivalent to the following:

$$\begin{aligned} x + y + z &= 0 \\ y + 1.5\,z &= 0 \end{aligned}$$

Setting $z = \theta$ as a parameter, this pair of equations can be written as

$$\begin{aligned} x + y &= -\theta \\ y &= -1.5\theta \end{aligned}$$

or

$$\begin{bmatrix} 1 & 1 \\ 0 & 1 \end{bmatrix} \begin{bmatrix} x \\ y \end{bmatrix} = \begin{bmatrix} -\theta \\ -1.5\theta \end{bmatrix} \qquad [3.34]$$

Premultiplying both sides by the inverse of $\begin{bmatrix} 1 & 1 \\ 0 & 1 \end{bmatrix}$, we get

$$\begin{aligned} \begin{bmatrix} x \\ y \end{bmatrix} &= \begin{bmatrix} 1 & 1 \\ 0 & 1 \end{bmatrix}^{-1} \begin{bmatrix} -\theta \\ -1.5\theta \end{bmatrix} \\ &= \begin{bmatrix} 1 & -1 \\ 0 & 1 \end{bmatrix} \begin{bmatrix} -\theta \\ -1.5\theta \end{bmatrix} \\ &= \begin{bmatrix} 0.5\theta \\ -1.5\theta \end{bmatrix} \end{aligned}$$

Therefore, the general solution is

$$\begin{bmatrix} x \\ y \\ z \end{bmatrix} = \begin{bmatrix} 0.5\theta \\ -1.5\theta \\ \theta \end{bmatrix} \qquad [3.35]$$

This case is illustrative of the general result that a system of m linear homogeneous equations $\mathbf{Ax} = \mathbf{0}$ involving n unknowns (variables) has an infinite number of solutions if the rank of $\mathbf{A}$ is less than n.

Notice that the parameter θ occurs multiplicatively on the right-hand side of 3.35. This means that if we set $\theta = 0$, we get the trivial solution. Another implication of the same feature is that each nonzero particular solution is a constant multiple of each other nonzero solution. For example, if we set $\theta = 2$ and $\theta = 4$, we get the following solutions

$$\begin{bmatrix} 1 \\ -3 \\ 2 \end{bmatrix} \quad \begin{bmatrix} 2 \\ -6 \\ 4 \end{bmatrix}$$

Clearly the second solution is twice the first. In general, in circumstances such as the one illustrated above, if $\mathbf{x}^* \neq \mathbf{0}$ is a solution of $\mathbf{Ax} = \mathbf{0}$, then any scalar multiple of $\mathbf{x}^*$ must also be a solution. (The circumstance under which this result holds is when the rank of the coefficient matrix is just one less than the number of unknowns involved.)

Consider now the following example:

$$\begin{aligned} x + y + u + v &= 0 \\ x - 3y - u - v &= 0 \\ 3x - y + u + v &= 0 \end{aligned} \qquad [3.36]$$

Here we have three equations in four unknowns (x,y,u,v). The coefficient matrix is

$$\begin{bmatrix} 1 & 1 & 1 & 1 \\ 1 & -3 & -1 & -1 \\ 3 & -1 & 1 & 1 \end{bmatrix}$$

with an echelon form

$$\begin{bmatrix} 1 & 1 & 1 & 1 \\ 0 & 1 & 1/2 & 1/2 \\ 0 & 0 & 0 & 0 \end{bmatrix}$$

telling us that the rank of the coefficient matrix is 2, and that the original equation system is equivalent to

$$\begin{aligned} x + y + u + v &= 0 \\ y + (1/2)u + (1/2)v &= 0 \end{aligned} \qquad [3.37]$$

Treating u and v as parameters (say, $u = \theta$ and $v = \phi$), we can write 3.37 as

$$\begin{bmatrix} 1 & 1 \\ 0 & 1 \end{bmatrix} \begin{bmatrix} x \\ y \end{bmatrix} = \begin{bmatrix} -(\theta + \phi) \\ -(1/2)\ (\theta + \phi) \end{bmatrix}$$

solving which we get

$$\begin{bmatrix} x \\ y \end{bmatrix} = \begin{bmatrix} 1 & -1 \\ 0 & 1 \end{bmatrix} \begin{bmatrix} -(\theta + \phi) \\ -(1.2)\ (\theta + \phi) \end{bmatrix}$$

$$= \begin{bmatrix} -(1/2)\ (\theta + \phi) \\ -(1/2)\ (\theta + \phi) \end{bmatrix}$$

The general solution we are looking for is therefore

$$\begin{bmatrix} x \\ y \\ u \\ v \end{bmatrix} = \begin{bmatrix} -(1/2)\ (\theta + \phi) \\ -(1/2)\ (\theta + \phi) \\ \theta \\ \phi \end{bmatrix} \qquad [3.38]$$

Notice that in this case each nonzero particular solution is not a scalar multiple of each other nonzero particular solution.

4. EIGENVALUES AND EIGENVECTORS

This chapter introduces the concepts of eigenvalues and eigenvectors. The application of these concepts in the principal component analysis is briefly discussed. To prepare the background, the concept of determinants is introduced first.

Determinants

The early development of determinants was connected with procedures for solving simultaneous equations. Consider the following two equations in two unknowns (x and y):

$$\begin{aligned} ax + by &= e \\ cx + dy &= f \end{aligned} \qquad [4.1]$$

Let us solve these equations by the method of elimination. Multiply the first equation by c and the second by a and then subract one from the other; we get $(ad - bc)y = af - ce$; hence, if $(ad - bc) \neq 0$,

$$y = (af - ce)/(ad - bc) \qquad [4.2]$$

Similarly, if we multiply the first equation in 4.1 by d and the second by b, and then subtract one from the other, we get $(ad - bc)x = de - bf$; hence, if $(ad - bc) \neq 0$,

$$x = (de - bf)/(ad - bc) \qquad [4.3]$$

Notice that 4.2 is a ratio of two quantities, the denominator being a number calculated from the elements of the coefficient matrix $\mathbf{A} = \begin{bmatrix} a & b \\ c & d \end{bmatrix}$ and the numerator a similar number calculated from the elements of $\begin{bmatrix} a & e \\ c & f \end{bmatrix}$. A parallel feature is characteristic of 4.3 also. This suggests the usefulness of introducing a procedure by which a unique scalar (ordinary number) is associated with a square matrix. The procedure should be such that its application to the coefficient matrix of 4.1 yields the scalar $(ad - bc)$ and must be generalizable to higher-order matrices. With this in mind, let us consider taking a weighted sum of the elements of the first or the second row of

$$\mathbf{A} = \begin{bmatrix} a & b \\ c & d \end{bmatrix}$$

such that in either case we get $(ad - bc)$. Obviously, in order to get $(ad - bc)$ as a weighted sum of a and b, we must weight a by d and b by $(-c)$, and to get the same result as a weighted sum of c and d, we must weight c by $(-b)$ and d by a. Let us put these suggested weights in matrix form as shown below

$$\mathbf{W} = \begin{bmatrix} d & -c \\ -b & a \end{bmatrix}$$

with the proposed weight for the (i,j) element of $\mathbf{A}$ appearing as the (i,j) element of $\mathbf{W}$. It so happens that if we omit the ith row and the jth column of $\mathbf{A}$ and then attach the sign $(-1)^{i+j}$ to the remaining element we get the (i,j) element of $\mathbf{W}$. Thus, if we remove the first row and second column of

A and attach to the remaining element the sign $(-1)^{1+2}$, we get $-c$, which is the (1,2) element of **W**. Also, we notice that

$$\mathbf{AW'} = \mathbf{W'A} = \begin{bmatrix} (ad - bc) & 0 \\ 0 & (ad - bc) \end{bmatrix}$$

which means that the weighted sum of the elements of any row (column) of **A** using as weights the corresponding elements of **W** is the unique scalar we are looking for, namely $(ad - bc)$. We call this quantity the *determinant* of **A**, written det(**A**), or more commonly $|\mathbf{A}|$, with vertical lines on either side. The (i,j) element of **W** is called the *cofactor* of the (i,j) element of **A**. This label is apt, since the reference is to a weighting factor in the computation of $|\mathbf{A}|$.

Let us stipulate that the determinant of a (1×1) matrix is the numerical value of the sole element of the matrix. The cofactor of the (i,j) element of **A** may now be defined as $(-1)^{i+j}$ times the determinant of the submatrix formed by omitting the ith row and the jth column of **A**. Having defined determinants of (2×2) matrices in terms of determinants of (1×1) matrices, it is possible to define the determinant of a (3×3) matrix in terms of determinants of (2×2) matrices as the weighted sum of the elements of any row or column of the given (3×3) matrix, using as weights the respective cofactors—the cofactor of the (i,j) element of the (3×3) matrix being $(-1)^{i+j}$ times the determinant of the (2×2) submatrix formed by omitting the ith row and j^{th} column of the original matrix. This definition is readily generalizable. A (3×3) case is examined below.

Consider the matrix

$$\mathbf{F} = \begin{bmatrix} 2 & 3 & 1 \\ 4 & 7 & 2 \\ 3 & 1 & 1 \end{bmatrix}$$

The following matrix, **G**, has for its (i,j) element the cofactor of the (i,j) element of **F**.

$$\mathbf{G} = \begin{bmatrix} 5 & 2 & -17 \\ -2 & -1 & 7 \\ -1 & 0 & 2 \end{bmatrix}$$

Notice that the (1,1) element of **G** is $(-1)^{1+1}$ times the determinant of the (2×2) submatrix $\begin{bmatrix} 7 & 2 \\ 1 & 1 \end{bmatrix}$ formed by omitting the first row and first column of **F**, the (3,2) element of **G** is $(-1)^{3+2}$ times the determinant of the submatrix formed by omitting the third row and second column of **F**, and so on.

It is easy to verify that

$$\mathbf{FG'} = \mathbf{G'F} = (-1)\mathbf{I}_3 \qquad [4.4]$$

demonstrating, among other things (see below), that the weighted sum of the elements of any row (column) of **F**, using as weights the respective cofactors, is the same as the weighted sum of the elements of any other row (column) similarly obtained, this unique number being (–1) in the present case. We thus have $|\mathbf{F}| = -1$.

Computer programs are available for the calculation of determinants. If one uses SAS, for example, the following statements will produce the determinant of **F** given above:

```
PROC MATRIX PRINT:
F = 2  3  1/4  7  2/3  1  1;
DELTA = DET(F):
```

The second statement sets up the matrix **F** (the slashes separating one row from the rest). The next statement calls for the calculation of |F| and for assigning the value to the variable DELTA.

Determinants do provide a method, albeit an inefficient one from the computational point of view, for inverting nonsingular, square matrices. The logic of the method rests on the interesting result that if $\mathbf{M}_{n\times n} = ((m_{ij}))$ is any $(n \times n)$ matrix and $\mathbf{C}_{n\times n} = ((c_{ij}))$ is such that c_{ij} is the cofactor of m_{ij} (for $i = 1, \ldots, n; j = 1, \ldots, n$), then

$$\mathbf{MC'} = \mathbf{C'M} = |\mathbf{M}|\mathbf{I}_n \qquad [4.5]$$

which we have already found to hold in a particular example (see 4.4 above). This implies, among other things (see below), that if we multiply each element of **C′** by the reciprocal of $|\mathbf{M}|$, provided, of course, $|\mathbf{M}| \neq 0$, the resulting matrix is $\mathbf{M}^{-1}$. Verify this using the matrix **F** examined above. Applying this technique to the matrix **A** given earlier (the coefficient matrix of 4.1), we get

$$\begin{bmatrix} a & b \\ c & d \end{bmatrix}^{-1} = \frac{1}{(ad - bc]} \begin{bmatrix} d & -b \\ -c & a \end{bmatrix}$$

if $(ad - bc) \neq 0$. This suggests the following easy-to-remember steps for inverting any (2×2) matrix that can be inverted:

Step 1. Calculate the determinant of the matrix. If the determinant is not equal to zero, go to the next step.

Step 2. Interchange the (1,1) element with the (2,2) element, and change the signs of the other elements.

Step 3. Divide each element of the matrix created in Step 2 by the determinant (if it is not equal to zero). The result is the inverse of the original matrix.

To illustrate, consider the matrix

$$\begin{bmatrix} 2 & 3 \\ 1 & 3 \end{bmatrix}$$

Its determinant is $(2 \times 3 - 1 \times 3) = 3 \neq 0$. Interchanging the (1,1) and (2,2) elements and changing the signs of the other elements of the given matrix, we get

$$\begin{bmatrix} 3 & -3 \\ -1 & 2 \end{bmatrix}$$

Dividing each element of this matrix by 3, the determinant of the original matrix, we get

$$\begin{bmatrix} 1 & -1 \\ -1/3 & 2/3 \end{bmatrix}$$

which is the inverse of the given matrix, as can be easily verified.

Note that if the determinant of a square matrix $\mathbf{M}_{n \times n} = ((m_{ij}))$ is zero, then from 4.5

$$\mathbf{MC}' = \mathbf{C}'\mathbf{M} = 0$$

where $\mathbf{C} = ((c_{ij}))$ is such that c_{ij} is the cofactor of $m_{ij} (i = 1, \ldots, n; j = 1, \ldots, n)$. This means that the rows (columns) of such matrices are linearly dependent. Such matrices, as we have already seen, cannot be inverted, and are called singular. We may now define a singular matrix as one whose determinant is zero. This is an alternative definition of singularity.

Determinants have several interesting properties, but we shall not examine them here. Interested readers may refer to Martin (1969), Noble (1969) and the references cited therein.

Eigenvalues and Eigenvectors

Associated with a square matrix are numbers called *eigenvalues* and vectors called *eigenvectors*. Consider the matrix

$$\mathbf{M} = \begin{bmatrix} 0.6 & 0.4 \\ 0.3 & 0.7 \end{bmatrix} \qquad [4.6]$$

An eigenvalue of **M** is any number λ such that $\mathbf{M}-\lambda\mathbf{I}$ is singular. In the present case it is easily verified that the numbers $\lambda_1 = 1$ and $\lambda_2 = 0.3$ have this property:

$$\mathbf{M} - \lambda_1\mathbf{I} = \begin{bmatrix} (0.6 - 1.0) & 0.4 \\ 0.3 & (0.7 - 1.0) \end{bmatrix} = \begin{bmatrix} -0.4 & 0.4 \\ 0.3 & -0.3 \end{bmatrix}$$

which is singular, and

$$\mathbf{M} - \lambda_2\mathbf{I} = \begin{bmatrix} (0.6 - 0.3) & 0.4 \\ 0.3 & (0.7 - 0.3) \end{bmatrix} = \begin{bmatrix} 0.3 & 0.4 \\ 0.3 & 0.4 \end{bmatrix}$$

which is also singular. Hence 1.0 and 0.3 are eigenvalues of

$$\begin{bmatrix} 0.6 & 0.4 \\ 0.3 & 0.7 \end{bmatrix}$$

To obtain the eigenvalues we solve the *determinantal* or *characteristic equation* $|\mathrm{M} - \lambda\mathrm{I}| = 0$, in which λ is treated as an unknown scalar. In the present case

$$|\mathrm{M} - \lambda\mathrm{I}| = \left| \begin{bmatrix} 0.6 & 0.4 \\ 0.3 & 0.7 \end{bmatrix} - \lambda \begin{bmatrix} 1 & 0 \\ 0 & 1 \end{bmatrix} \right|$$

$$= \begin{vmatrix} 0.6 - \lambda & 0.4 \\ 0.3 & 0.7 - \lambda \end{vmatrix}$$

Hence the determinantal equation just mentioned is

$$\begin{vmatrix} 0.6 - \lambda & 0.4 \\ 0.3 & 0.7 - \lambda \end{vmatrix} = 0$$

which is the same as

$$(0.6-\lambda)(0.7-\lambda) - (0.3)(0.4) = 0$$

or on simplification,

$$\lambda^2 - (1.3)(\lambda) + (0.3) = 0 \qquad [4.7]$$

This equation, being a quadratic, has two and only two roots: $\lambda_1 = 1.0$ and $\lambda_2 = 0.3$, which are the eigenvalues (or sometimes called the *latent roots* or the *characteristic roots*) of the original matrix **M**. In the case of a 3×3 matrix, the determinantal equation is a cubic; hence it has three roots (eigenvalues). In general, an $n \times n$ matrix has n eigenvalues. Sometimes some of the eigenvalues may be repeated. Thus, for the matrix

$$\begin{bmatrix} 3 & 0 \\ 0 & 3 \end{bmatrix}$$

the eigenvalues being the roots of the equation $(3 - \lambda)^2 = 0$ are repeated (= 3); we say $\lambda = 3$ is of multiplicity 2 in this instance.

In some instances the eigenvalues may not be real-valued. For example, in the case of

$$\mathbf{R} = \begin{bmatrix} 1 & -3 \\ 4 & 1 \end{bmatrix}$$

the characteristic equation is

$$\begin{vmatrix} 1-\lambda & -3 \\ 4 & 1-\lambda \end{vmatrix} = 0$$

that is,

$$(1 - \lambda)^2 + 12 = 0$$

giving $\lambda_1 = 1 + \sqrt{-12}$ and $\lambda_2 = 1 - \sqrt{-12}$ as the eigenvalues, which are obviously not real-valued, as the presence of the square root of a negative number signifies.

It should be emphasized that solving characteristic equations is not always a simple matter. Fortunately, computer programs are available for the purpose. If one uses the Matrix Procedure of SAS, for example, the matrix is specified and its eigenvalues are created, using the following statements:

```
PROC MATRIX PRINT;
A = 1  2/4  3;
M = EIGVAL (A);
```

The second statement specifies the matrix, using the slash to separate one row from another. The next statement calls for the eigenvalues. The printout shows the eigenvalues as a vector **M**.

If λ_1 is an eigenvalue of a given matrix **A**, it is possible to find a vector **c** such that

$$\mathbf{Ac} = \lambda_1 \mathbf{c} \qquad [4.8]$$

Any **c** that satisfies equation 4.8 is called an eigenvector associated with the eigenvalue in question ($= \lambda_1$). If **c** is an eigenvector, so is $k\mathbf{c}$, where k is a scalar.

For the matrix **M** given in [4.6], an eigenvector associated with the eigenvalue 0.3 can be computed by solving the following equations for the unknowns c_1 and c_2:

$$\begin{bmatrix} 0.6 & 0.4 \\ 0.3 & 0.7 \end{bmatrix} \begin{bmatrix} c_1 \\ c_2 \end{bmatrix} = (0.3) \begin{bmatrix} c_1 \\ c_2 \end{bmatrix} \qquad [4.9]$$

This system of equations is the same as

$$\begin{aligned} (0.6)(c_1) + (0.4)(c_2) &= (0.3)(c_1) \\ (0.3)(c_1) + (0.7)(c_2) &= (0.3)(c_1) \end{aligned}$$

When the terms on the right-hand side are brought to the left-hand side, these equations become

$$\begin{aligned} (0.3)(c_1) + (0.4)(c_2) &= 0 \\ (0.3)(c_1) + (0.4)(c_2) &= 0 \end{aligned} \qquad [4.10]$$

Immediately we notice that we have a system of homogeneous equations (in two unknowns) and that the rank of the coefficient matrix is 1, which is one less than the number of unknowns. Put differently, we have only one equation connecting the two unknowns c_1 and c_2; an infinite number of solutions exist, each a constant multiple of the others; and the general solution is (treating c_1 as a parameter)

$$\begin{aligned} c_1 &= \theta \\ c_2 &= -0.75\theta \end{aligned} \qquad [4.11]$$

θ being arbitrary (any number, positive, negative, or zero). One particular solution is of special interest, and that is the one for which the sum of squares of the values is equal to unity. In the present case it is easily seen that this particular solution is

$$c_1 = 0.8$$
$$c_2 = -0.6 \qquad [4.12]$$

(To obtain these, set θ equal to any non-zero value in 4.11, then divide each value by the square root of the sum of squares of the values. Thus, if θ is set equal to 1, the two values are $c_1 = 1$ and $c_2 = -0.75$; the sum of the squares of these values is 1.5625, the square root of which is 1.25; division now gives $c_1 = 1/1.25 = 0.8$ and $c_2 = (-0.75)/1.25 = -0.6$.)

A particular form of 4.9 is thus

$$\begin{bmatrix} 0.6 & 0.4 \\ 0.3 & 0.7 \end{bmatrix} \begin{bmatrix} 0.8 \\ -0.6 \end{bmatrix} = (0.3) \begin{bmatrix} 0.8 \\ -0.6 \end{bmatrix} \qquad [4.13]$$

(Verify by multiplication that this indeed holds.) Corresponding to the eigenvalue 1.0 of **M**, we similarly obtain the eigenvector

$$\begin{bmatrix} 1/\sqrt{2} \\ 1/\sqrt{2} \end{bmatrix}$$

which satisfies the equation

$$\begin{bmatrix} 0.6 & 0.4 \\ 0.3 & 0.7 \end{bmatrix} \begin{bmatrix} 1/\sqrt{2} \\ 1/\sqrt{2} \end{bmatrix} = (1.0) \begin{bmatrix} 1/\sqrt{2} \\ 1/\sqrt{2} \end{bmatrix} \qquad [4.14]$$

Putting 4.13 and 4.14 together we get

$$\begin{bmatrix} 0.6 & 0.4 \\ 0.3 & 0.7 \end{bmatrix} \begin{bmatrix} 1/\sqrt{2} & 0.8 \\ 1/\sqrt{2} & -0.6 \end{bmatrix} = \begin{bmatrix} 1/\sqrt{2} & 0.8 \\ 1/\sqrt{2} & -0.6 \end{bmatrix} \begin{bmatrix} 1.0 & 0.0 \\ 0.0 & 0.3 \end{bmatrix} \qquad [4.15]$$

Postmultiplying both sides of 4.15 by

$$\begin{bmatrix} 1/\sqrt{2} & 0.8 \\ 1/\sqrt{2} & -0.6 \end{bmatrix}^{-1}$$

gives

$$\begin{bmatrix} 0.6 & 0.4 \\ 0.3 & 0.7 \end{bmatrix} = \begin{bmatrix} 1/\sqrt{2} & 0.8 \\ 1/\sqrt{2} & -0.6 \end{bmatrix} \begin{bmatrix} 1.0 & 0.0 \\ 0.0 & 0.3 \end{bmatrix} \begin{bmatrix} 1/\sqrt{2} & 0.8 \\ 1/\sqrt{2} & -0.6 \end{bmatrix}^{-1} \quad [4.16]$$

which expresses the original matrix **M** as a product $\mathbf{P}\Lambda\mathbf{P}^{-1}$ where Λ is a diagonal matrix (i.e., one with nonzero elements only in the principal diagonal). Expressing a (square) matrix in this form is known as *diagonalization.*

Not all square matrices can be diagonalized. An example will make this clear. Consider the matrix

$$\mathbf{S} = \begin{bmatrix} 3 & 0 \\ 4 & 3 \end{bmatrix}$$

This matrix has $\lambda_1 = 3$ and $\lambda_2 = 3$ as its eigenvalues. (Notice that the eigenvalue is repeated.) To obtain an eigenvector corresponding to $\lambda_1 = 3$ we solve for c_1 and c_2 from

$$\begin{bmatrix} 3 & 0 \\ 4 & 3 \end{bmatrix} \begin{bmatrix} c_1 \\ c_2 \end{bmatrix} = 3 \begin{bmatrix} c_1 \\ c_2 \end{bmatrix}$$

that is, from

$$3c_1 = 3c_1$$
$$4c_1 + 3c_2 = 3c_2$$

The solution is $c_1 = 0$, $c_2 = \theta$, where θ is arbitrary. The eigenvector of unit norm (i.e., one with the sum of squares of its elements equal to unity) corresponding to $\lambda_1 = 3$ is thus

$$\mathbf{p}_1 = \begin{bmatrix} 0 \\ 1 \end{bmatrix}$$

(Of course, one could choose $-\mathbf{p}_1$ instead.) We now search for an eigenvector of unit norm corresponding to $\lambda_2 = 3$. Obviously, $\mathbf{p}_1$ obtained above (or $-\mathbf{p}_1$) qualifies to be an eigenvector associated with $\lambda_2 = 3$. If we associate with both $\lambda_1 = 3$ and $\lambda_2 = 3$ the same eigenvector, say, $\mathbf{p}_1$, obtained above, we get

$$\begin{bmatrix} 3 & 0 \\ 4 & 3 \end{bmatrix} \begin{bmatrix} 0 & 0 \\ 1 & 1 \end{bmatrix} = \begin{bmatrix} 0 & 0 \\ 1 & 1 \end{bmatrix} \begin{bmatrix} 3 & 0 \\ 0 & 3 \end{bmatrix}$$

which is of the form

$$\mathbf{SP} = \mathbf{P\Lambda}$$

where $\mathbf{P} = \begin{bmatrix} 0 & 0 \\ 1 & 1 \end{bmatrix}$ and $\Lambda = \begin{bmatrix} 3 & 0 \\ 0 & 3 \end{bmatrix}$. But in this case $\mathbf{P}^{-1}$ does not exist and consequently $\mathbf{S}$ cannot be expressed in the form $\mathbf{S} = \mathbf{P\Lambda P}^{-1}$. The same is true if we associate $\mathbf{p}_1$ with λ_1 and $-\mathbf{p}_1$ with λ_2, or vice versa. So it is unproductive to associate with $\lambda_1 = 3$ and $\lambda_2 = 3$ eigenvectors that are multiples of each other. We should look for an eigenvector corresponding to $\lambda_2 = 3$, which is not a multiple of the eigenvector already associated with $\lambda_1 = 3$. This means that we should search for d_1 and d_2 that satisfy all of the following conditions:

$$\begin{bmatrix} 3 & 0 \\ 4 & 3 \end{bmatrix} \begin{bmatrix} d_1 \\ d_2 \end{bmatrix} = \lambda_2 \begin{bmatrix} d_1 \\ d_2 \end{bmatrix} \quad [4.17]$$

where $\lambda_2 = 3$;

$$\begin{bmatrix} d_1 \\ d_2 \end{bmatrix} \neq k \begin{bmatrix} 0 \\ 1 \end{bmatrix} \quad [4.18]$$

whatever be k; and

$$d_1^2 + d_2^2 = 1 \quad [4.19]$$

The only d_1 and d_2 that satisfy 4.17 are $d_1 = 0$ and $d_2 = \theta$, where θ is arbitrary. If the chosen d_1 and d_2 must also satisfy 4.19, θ must be 1 (or -1). This leads to

$$\begin{bmatrix} 0 \\ 1 \end{bmatrix} \text{ or } \begin{bmatrix} 0 \\ -1 \end{bmatrix}$$

as our choice for the eigenvector to be associated with $\lambda_2 = 3$. But neither of these satisfies 4.18. In fact it is impossible to find d_1 and d_2 that satisfy all the conditions laid out above. We thus have a case in which it is impossible to find a $\mathbf{P}$ such that $\mathbf{S}$ can be expressed in the form $\mathbf{P\Lambda P}^{-1}$.

Another important point worth noting is that some matrices with repeated eigenvalues have an infinite number of diagonal forms. To give an example, consider the matrix

$$\mathbf{X} = \begin{bmatrix} 4 & -1 & -1 \\ -1 & 4 & 1 \\ -1 & 1 & 4 \end{bmatrix}$$

It is not difficult to verify that $\lambda_1 = 6$, $\lambda_2 = 3$, and $\lambda_3 = 3$ are the eigenvalues of this matrix. (Notice that 3 is of multiplicity 2.) To find an eigenvector associated with $\lambda_1 = 6$, we solve for c_1, c_2, and c_3 from

$$\begin{bmatrix} 4 & -1 & -1 \\ -1 & 4 & 1 \\ -1 & 1 & 4 \end{bmatrix} \begin{bmatrix} c_1 \\ c_2 \\ c_3 \end{bmatrix} = 6 \begin{bmatrix} c_1 \\ c_2 \\ c_3 \end{bmatrix}$$

that is, from

$$\begin{bmatrix} -2 & -1 & -1 \\ -1 & -2 & 1 \\ -1 & 1 & -2 \end{bmatrix} \begin{bmatrix} c_1 \\ c_2 \\ c_3 \end{bmatrix} = \begin{bmatrix} 0 \\ 0 \\ 0 \end{bmatrix}$$

The solution of this system is

$$\mathbf{v}_1 = \begin{bmatrix} c_1 \\ c_2 \\ c_3 \end{bmatrix} = \begin{bmatrix} -\theta \\ \theta \\ \theta \end{bmatrix} \qquad [4.20]$$

where θ is arbitrary. To get an eigenvector of unit norm we divide each element of $\mathbf{v}_1$ by the square root of the sum of squares of its elements (after giving θ a nonzero value). We get by this process

$$\mathbf{p}_1 = \begin{bmatrix} -1/\sqrt{3} \\ 1/\sqrt{3} \\ 1/\sqrt{3} \end{bmatrix} \qquad [4.21]$$

(Notice that $-\mathbf{p}_1$ also would be appropriate.) We now search for an eigenvector associated with $\lambda_2 = 3$, by solving for d_1, d_2, and d_3 from

$$\begin{bmatrix} 4 & -1 & -1 \\ -1 & 4 & 1 \\ -1 & 1 & 4 \end{bmatrix} \begin{bmatrix} d_1 \\ d_2 \\ d_3 \end{bmatrix} = 3 \begin{bmatrix} d_1 \\ d_2 \\ d_3 \end{bmatrix} \qquad [4.22]$$

that is, from

$$\begin{bmatrix} 1 & -1 & -1 \\ -1 & 1 & 1 \\ -1 & 1 & 1 \end{bmatrix} \begin{bmatrix} d_1 \\ d_2 \\ d_3 \end{bmatrix} = \begin{bmatrix} 0 \\ 0 \\ 0 \end{bmatrix}$$

the solution of this system is of the form

$$\begin{bmatrix} d_1 \\ d_2 \\ d_3 \end{bmatrix} = \begin{bmatrix} \delta + \theta \\ \delta \\ \theta \end{bmatrix}$$

where δ and θ are both arbitrary. Not that we have a "doubly" infinite number of choices here; we may give any value for δ and any value for θ. Suppose we give δ the value 0; this leads to the following vector of unit norm

$$\mathbf{p}_2 = \begin{vmatrix} 1/\sqrt{2} \\ 0 \\ 1/\sqrt{2} \end{vmatrix} \qquad [4.23]$$

(It is, of course, all right to choose $-\mathbf{p}_2$ instead). To get an eigenvector associated with $\lambda_3 = 3$ we proceed in the same way as was done in the case of $\lambda_2 = 3$, and find that the choice is of the form

$$\begin{bmatrix} e_1 \\ e_2 \\ e_3 \end{bmatrix} = \begin{bmatrix} \delta + \theta \\ \delta \\ \theta \end{bmatrix} \qquad [4.24]$$

where δ and θ are arbitrary. We may choose any vector of this form that is not a constant multiple of $\mathbf{p}_2$ already chosen. Thus one choice may be to set θ equal to zero. This leads to

$$\mathbf{p}_3 = \begin{bmatrix} 1/\sqrt{2} \\ 1/\sqrt{2} \\ 0 \end{bmatrix}$$

We thus have the following as one of the many possible diagonalizations of $\mathbf{X}$:

$$\mathbf{X} = \mathbf{P}\Lambda\mathbf{P}^{-1}$$

where

$$\mathbf{P} = \begin{bmatrix} -1/\sqrt{3} & 1/\sqrt{2} & 1/\sqrt{2} \\ 1/\sqrt{3} & 0 & 1/\sqrt{2} \\ 1/\sqrt{3} & 1/\sqrt{2} & 0 \end{bmatrix} \quad \text{and } \Lambda = \begin{bmatrix} 6 & 0 & 0 \\ 0 & 3 & 0 \\ 0 & 0 & 3 \end{bmatrix}$$

Verify that

$$\mathbf{P}^{-1} = \begin{bmatrix} -1/\sqrt{3} & 1/\sqrt{3} & 1/\sqrt{3} \\ \sqrt{2/3} & -\sqrt{2/3} & 2\sqrt{2/3} \\ \sqrt{2/3} & 2\sqrt{2/3} & -\sqrt{2/3} \end{bmatrix}$$

and that $\mathbf{P\Lambda P}^{-1}$ indeed reproduces $\mathbf{X}$.

It may be of interest to note in passing that

$$\begin{bmatrix} 1/\sqrt{6} \\ 2/\sqrt{6} \\ -1/\sqrt{6} \end{bmatrix} \qquad [4.25]$$

is also a possible choice for an eigenvector of unit norm corresponding to $\lambda_3 = 3$. (This vector was arrived at by requiring, to begin with, that when 4.24 is premultiplied by $\mathbf{p}_2'$ from 4.23 we get 0. The motivation for doing so will become clear later. The vector thus obtained when transformed to one of unit norm yielded 4.25.) With 4.21, 4.23, and 4.25 the diagonalization becomes $\mathbf{X} = \mathbf{Q\Lambda Q'}$ where

$$\mathbf{Q} = \begin{bmatrix} -1/\sqrt{3} & 1/\sqrt{2} & 1/\sqrt{6} \\ 1/\sqrt{3} & 0 & 2/\sqrt{6} \\ 1/\sqrt{3} & 1/\sqrt{2} & -1/\sqrt{6} \end{bmatrix}$$

and $\mathbf{QQ'} = \mathbf{I}$. We return in section 4.4 to matrices that can be diagonalized in this latter fashion. We close this section by pointing out that diagonalization of a matrix facilitates obtaining higher powers of the matrix.

Thus, if

$$\mathbf{M} = \mathbf{P\Lambda P}^{-1}$$

we notice that

$$\begin{aligned} \mathbf{M}^2 = \mathbf{MM} &= (\mathbf{P\Lambda P}^{-1})(\mathbf{P\Lambda P}^{-1}) \\ &= \mathbf{P\Lambda}(\mathbf{P}^{-1}\mathbf{P})\mathbf{\Lambda P}^{-1} \\ &= \mathbf{P\Lambda}^2\mathbf{P}^{-1} \end{aligned}$$

and similarly

$$\mathbf{M}^3 = \mathbf{P\Lambda}^3\mathbf{P}^{-1}$$

and in general

$$\mathbf{M}^n = \mathbf{P}\Lambda^n\mathbf{P}^{-1}$$

The attractiveness of this result is that since Λ is diagonal we can find its powers easily (by simply raising the diagonal elements to the desired power).

From 4.16 we thus have

$$\begin{bmatrix} 0.6 & 0.4 \\ 0.3 & 0.7 \end{bmatrix}^n = \begin{bmatrix} 1/\sqrt{2} & 0.8 \\ 1/\sqrt{2} & -0.6 \end{bmatrix} \begin{bmatrix} 1.0 & 0.0 \\ 0.0 & (0.3)^n \end{bmatrix} \begin{bmatrix} 1/\sqrt{2} & 0.8 \\ 1/\sqrt{2} & -0.6 \end{bmatrix}^{-1}$$

For large n we know that $(0.3)^n$ is practically equal to zero. Hence, for large n

$$\begin{bmatrix} 0.6 & 0.4 \\ 0.3 & 0.7 \end{bmatrix}^n = \begin{bmatrix} 1/\sqrt{2} & 0.8 \\ 1/\sqrt{2} & -0.6 \end{bmatrix} \begin{bmatrix} 1.0 & 0.0 \\ 0.0 & 0.0 \end{bmatrix} \begin{bmatrix} 1/\sqrt{2} & 0.8 \\ 1/\sqrt{2} & -0.6 \end{bmatrix}^{-1}$$

$$= \begin{bmatrix} 1/\sqrt{2} & 0.0 \\ 1/\sqrt{2} & 0.0 \end{bmatrix} \begin{bmatrix} 1/\sqrt{2} & 0.8 \\ 1/\sqrt{2} & -0.6 \end{bmatrix}^{-1}$$

$$= \begin{bmatrix} 3/7 & 4/7 \\ 3/7 & 4/7 \end{bmatrix}$$

Principal Components

Principal component analysis is a multivariate technique for examining relationships among several quantitative variables. It is often used for summarizing data. Thus given, for each county in the United States, data on per capita income, median years of education, percentage unemployed, homicide rate, percentage of women in correctional institutions, and so on, the analyst might be interested in exploring the possibility of summarizing the given data in terms of as few linear combinations of the data as possible with as little sacrifice of information as possible.

Given a data set with p quantitative variables, p principal components may be computed, each of which is a linear combination of the original variables, using as weights the elements of the eigenvectors of the correlation or covariance matrix or the matrix of sum of squares and products (corrected for the mean). The eigenvectors are customarily

TABLE 4.1

Person	Test 1	Test 2	Person	Test 1	Test 2
1	0.9	0.2	7	0.0	−0.1
2	0.8	0.4	8	−0.2	−0.7
3	0.5	0.4	9	−0.2	−0.6
4	0.2	0.6	10	−0.5	−0.4
5	0.2	0.7	11	−0.8	−0.4
6	0.0	0.1	12	−0.9	−0.2

taken with *unit-norm* (i.e., with the sum of squares of elements equaling unity). The principle behind the method can be illustrated in a simple example.

Suppose we are given the data shown in Table 4.1. Imagine that these are test scores, of 12 persons on two tests, expressed as deviations from the respective means.

The matrix of sums of squares and products computed from these figures is

$$\begin{array}{cc} \text{Test 1} & \text{Test 2} \end{array}$$

$$\mathbf{W} = \begin{bmatrix} 3.56 & 1.92 \\ 1.92 & 2.44 \end{bmatrix} \begin{array}{l} \text{Test 1} \\ \text{Test 2} \end{array}$$

The first step in the principal component analysis is the computation of the eigenvalues and their associated eigenvectors of the correlation matrix, the variance-covariance matrix, or, very infrequently, of the matrix of sums of squares and products. I will focus on the last-mentioned. The corresponding results for the variance-covariance matrix can be easily obtained from those based on the sums of squares and products.

The determinantal equation giving the eigenvalues of $\mathbf{W}$ is $|\mathbf{W} - \lambda\mathbf{I}| = 0$, that is

$$\begin{vmatrix} 3.56 - \lambda & 1.92 \\ 1.92 & 2.44 - \lambda \end{vmatrix} = 0 \qquad [4.26]$$

which is the same as

$$(3.56 - \lambda)(2.44 - \lambda) - (1.92)^2 = 0$$

or, on simplification,

$$\lambda^2 - 6\lambda + 5 = 0 \qquad [4.27]$$

Solving this quadratic equation, we get $\lambda_1 = 5$ and $\lambda_2 = 1$ as the eigenvalues. Now, to obtain the eigenvector associated with the eigenvalue 5, we solve the equation system

$$\begin{bmatrix} 3.56 & 1.92 \\ 1.92 & 2.44 \end{bmatrix} \begin{bmatrix} c_1 \\ c_2 \end{bmatrix} = 5 \begin{bmatrix} c_1 \\ c_2 \end{bmatrix} \qquad [4.28]$$

which is equivalent to

$$\begin{aligned} -1.44\, c_1 + 1.92\, c_2 &= 0 \\ 1.92\, c_1 - 2.56\, c_2 &= 0 \end{aligned} \qquad [4.29]$$

The coefficient matrix in 4.29 is of rank 1, hence the number of solutions is infinite. [Verify that both of the equations in 4.29 are equivalent to $(-0.75)\, c_1 + c_2 = 0$.] The general solution is, treating c_1 as a parameter,

$$\begin{aligned} c_1 &= \theta \\ c_2 &= 0.75\theta \end{aligned}$$

and the particular solution with unit-norm is

$$\begin{aligned} c_1 &= 0.8 \\ c_2 &= 0.6 \end{aligned} \qquad [4.30]$$

Similarly, corresponding to the eigenvalue 1 we have the eigenvector $\begin{bmatrix} d_1 \\ d_2 \end{bmatrix}$ given by the solution to the equation

$$\begin{bmatrix} 3.56 & 1.92 \\ 1.92 & 2.44 \end{bmatrix} \begin{bmatrix} d_1 \\ d_2 \end{bmatrix} = (1) \begin{bmatrix} d_1 \\ d_2 \end{bmatrix} \qquad [4.31]$$

This equation system is equivalent to

$$\begin{aligned} 2.56\, d_1 + 1.92\, d_2 &= 0 \\ 1.92\, d_1 + 1.44\, d_2 &= 0 \end{aligned} \qquad [4.32]$$

[or $(4/3)d_1 + d_2 = 0$ repeated], with the general solution (treating d_1 as a parameter)

$$d_1 = \theta$$
$$d_2 = -(4/3)\,\theta$$

leading to the solution with unit-norm

$$d_1 = 0.6$$
$$d_2 = -0.8 \qquad [4.33]$$

The eigenvector 4.30 provides the weights for forming one linear combination of the original variables, and the eigenvector 4.33 provides another set of weights. We can thus form the following linear combinations of the test scores of person 1:

$$[0.9 \quad 0.2] \begin{bmatrix} 0.8 \\ 0.6 \end{bmatrix}$$

$$[0.9 \quad 0.2] \begin{bmatrix} 0.6 \\ -0.8 \end{bmatrix}$$

The former is the score on the first principal component and the latter the score on the second principal component. The scores thus calculated for all the cases in Table 4.1 are shown in Table 4.2.

Notice that the sum of squares (corrected for the mean = 0) of the scores on component 1 is 5, and that the sum of squares of the scores on component 2 is 1, these being equal to the corresponding eigenvalues.

The strategy usually followed in the principal component analysis is, as was done above, to sort the principal components by descending order of the eigenvalues. This is to help decide which components collectively capture most of the information in the data. In the illustrative material presented above, we may, for example, declare that the given data can be effectively summarized in terms of the first component, since the corresponding eigenvalue is five-sixths of the sum of the two eigenvalues. The following model for the data lies behind this approach:

$$\mathbf{Y} = \mathbf{XB} + \mathbf{E} \qquad [4.34]$$

where (speaking in general terms)

Y is an $n \times p$ matrix of centered, observed variables ($n = 12$ and $p = 2$ in our illustrative case);

X is the $n \times k$ matrix of scores on the first k principal components (in the illustrative case $k = 1$, and **X** is the vector in column 2 of Table 4.2);

B is the $k \times p$ matrix of eigenvectors (in our illustrative case **B** = [0.8 0.6], the eigenvector corresponding to the larger of the two eigenvalues); and

E is an $n \times p$ matrix of residuals (in our illustrative case **E** is a 12×2 matrix).

It is possible to show that the first k principal components give a least-squares fit of the model 4.34, minimizing the sum of squares of all the elements in **E**. For the illustrative material presented above, the various parts of 4.34 are as shown below:

$$\begin{bmatrix} 0.9 & 0.2 \\ 0.8 & 0.4 \\ 0.5 & 0.4 \\ 0.2 & 0.6 \\ 0.2 & 0.7 \\ 0.0 & 0.1 \\ 0.0 & -0.1 \\ -0.2 & -0.7 \\ -0.2 & -0.6 \\ -0.5 & -0.4 \\ -0.8 & -0.4 \\ -0.9 & -0.2 \end{bmatrix} = \begin{bmatrix} 0.672 & 0.504 \\ 0.704 & 0.528 \\ 0.512 & 0.384 \\ 0.416 & 0.312 \\ 0.464 & 0.348 \\ 0.048 & 0.036 \\ -0.048 & -0.036 \\ -0.464 & -0.348 \\ -0.416 & -0.312 \\ -0.512 & -0.384 \\ -0.704 & -0.528 \\ -0.672 & -0.504 \end{bmatrix} + \begin{bmatrix} 0.228 & -0.304 \\ 0.096 & -0.128 \\ -0.012 & 0.016 \\ -0.216 & 0.288 \\ -0.264 & 0.352 \\ -0.048 & 0.064 \\ 0.048 & -0.064 \\ 0.264 & -0.352 \\ 0.216 & -0.288 \\ 0.012 & -0.016 \\ -0.096 & 0.128 \\ -0.228 & 0.304 \end{bmatrix} \quad [4.35]$$

Verify that the first matrix on the right of 4.35 results from postmultiplying the column vector of scores on component 1 (Table 4.2) by [0.8 0.6], and that the second matrix on the right results from postmultiplying the column vector of scores on component 2 by [0.6 −0.8]. Verify also the following for the representation in 4.35.

—The sum of squares of the elements in the first matrix on the right is equal to 5, and the sum of squares of the elements in the second matrix on the right is equal to 1.

—When the two sums of squares mentioned above are added together the result equals the sum of squares of the elements in the matrix on the left.

TABLE 4.2

Person	*Component 1*	*Component 2*
1	0.84	0.38
2	0.88	0.16
3	0.64	−0.02
4	0.52	−0.36
5	0.58	−0.44
6	0.06	−0.08
7	−0.06	0.08
8	−0.58	0.44
9	−0.52	0.36
10	−0.64	0.02
11	−0.88	−0.16
12	−0.84	−0.38
Sum of squares	5.00	1.00
Sum of products	0.0	

The decomposition 4.35 suggests that we may think of the principal component method as one that decomposes a given matrix of centered observations into two parts, one of which pertains to k of the larger eigenvalues of the matrix of sum of squares and products and the other pertains to the remaining smaller eigenvalues.

Symmetric Matrices

In the preceding section, as we discussed the matrix of sums of squares and products of observations on two or more variables, we failed to take particular note of the point that such matrices are symmetric in that their (i,j) element is equal to their (j,i) element, for all i and j. Obviously this notion applies only to square matrices. We do come across symmetric matrices frequently in our research. The variance-covariance matrix, the correlation matrix, and the identity matrix of any order are examples of symmetric matrices.

Symmetric matrices whose elements are all real numbers (as distinguished from imaginary or complex numbers) are called *real symmetric matrices*. A few of their properties are mentioned below.

In the preceding section we saw that the real symmetric matrix

$$\mathbf{W} = \begin{bmatrix} 3.56 & 1.92 \\ 1.92 & 2.44 \end{bmatrix}$$

has $\lambda_1 = 5$ and $\lambda_2 = 1$ as its eigenvalues and that $\begin{bmatrix} 0.8 \\ 0.6 \end{bmatrix}$ and $\begin{bmatrix} 0.6 \\ -0.8 \end{bmatrix}$ are eigenvectors with unit-norm corresponding to these eigenvalues. To make it more interesting, let us use instead of $\begin{bmatrix} 0.6 \\ -0.8 \end{bmatrix}$ its negative, i.e., $\begin{bmatrix} -0.6 \\ 0.8 \end{bmatrix}$ as the eigenvector associated with $\lambda_2 = 1$. Diagonalization of **W** using these eigenvalues and eigenvectors gives

$$\mathbf{W} = \begin{bmatrix} 0.8 & -0.6 \\ 0.6 & 0.8 \end{bmatrix} \begin{bmatrix} 5 & 0 \\ 0 & 1 \end{bmatrix} \begin{bmatrix} 0.8 & -0\ 6 \\ 0.6 & 0.8 \end{bmatrix}^{-1}$$

$$= \begin{bmatrix} 0.8 & -0.6 \\ 0.6 & 0.8 \end{bmatrix} \begin{bmatrix} 5 & 0 \\ 0 & 1 \end{bmatrix} \begin{bmatrix} 0.8 & 0\ 6 \\ -0.6 & 0.8 \end{bmatrix}$$

which is of the form $\mathbf{P\Lambda P'}$, where Λ is a diagonal matrix and **P** satisfies the relationship $\mathbf{PP'} = \mathbf{P'P} = \mathbf{I}$. This result holds in general for real symmetric matrices.

If **A** is an $n \times n$ real symmetric matrix with eigenvalues $\lambda_1, \ldots, \lambda_n$, including repetitions (multiplicities), there exists a matrix **P** such that $\mathbf{A} = \mathbf{P\Lambda P'}$, where Λ is a diagonal matrix whose i^{th} diagonal element is λ_i and an eigenvector corresponding to λ_i forms the ith column of **P**, with **P** satisfying the relationship $\mathbf{PP'} = \mathbf{P'P} = \mathbf{I}$. (In passing, we note that any matrix **P** that satisfies the relationship $\mathbf{PP'} = \mathbf{P'P} = \mathbf{I}$ is called an *orthogonal* matrix.)

It can be shown that real symmetric matrices have only real numbers for their eigenvalues and that corresponding to any eigenvalue of a real symmetric matrix, it is possible to find a real eigenvector.

Furthermore, it can be shown that real symmetric matrices of the form $\mathbf{X'X}$ (e.g., the matrix of sums of squares and products) have only real nonnegative numbers for their eigenvalues. You might have guessed that this should be so, when the correspondence between the eigenvalues of **W** and the sums of squares of scores on principal components was mentioned in the preceding section.

Symmetric matrices have several other interesting properties, but space limitation prevents any discussion of them here.

REFERENCES

BROWN, E. K. (1967) Mathematics with Applications in Management and Economics. Homewood, IL: Irwin.

HADLEY, G. (1961) Linear Algebra. Reading, MA: Addison-Wesley.

KEMENY, G., A. SCHLEIFER, J. L. SNELL, AND G. L. THOMPSON (1962) Finite Mathematics with Business Applications. Englewood Cliffs, NJ: Prentice-Hall.

NOBLE, B. (1969) Applied Linear Algebra. Englewood Cliffs, NJ: Prentice-Hall.

MARTIN, E. W., Jr. (1969) Mathematics for Decision Making, Vol. 1: Linear Mathematics (Book 4: Vectors and Matrices). Homewood, IL: Irwin.

MILLS, G. (1969) Introduction to Linear Algebra: A Primer for Social Scientists. Chicago: Aldine.

SEARL, S. R. (1982) Matrix Algebra Useful for Statistics. New York: John Wiley.

KRISHNAN NAMBOODIRI is Professor of Sociology at the University of North Carolina at Chapel Hill, where he chaired the Department of Sociology from 1975-1980. He also served as editor of Demography, *1975-1978. His publications include* Applied Multivariate Analysis *and* Experimental Designs *(joint author). He is a Fellow of the American Statistical Association.*

Quantitative Applications in the Social Sciences

A SAGE UNIVERSITY PAPER SERIES

This series of methodological works provides introductory explanations and demonstrations of various data analysis techniques applicable to the social sciences. Designed for readers with a limited background in statistics or mathematics, this series aims to make the assumptions and practices of quantitative analysis more readily accessible.

1. **ANALYSIS OF VARIANCE** by Gudmund R. Iversen and Helmut Norpoth
2. **OPERATIONS RESEARCH METHODS** by Stuart Nagel with Marian Neef
3. **CAUSAL MODELING**, Second Edition, by Herbert B. Asher
4. **TESTS OF SIGNIFICANCE** by Ramon E. Henkel
5. **COHORT ANALYSIS** by Norval D. Glenn
6. **CANONICAL ANALYSIS AND FACTOR COMPARISON** by Mark S. Levine
7. **ANALYSIS OF NOMINAL DATA**, Second Edition, by H. T. Reynolds
8. **ANALYSIS OF ORDINAL DATA** by David K. Hildebrand, James D. Laing, and Howard Rosenthal
9. **TIME SERIES ANALYSIS: Regression Techniques** by Charles W. Ostrom, Jr.
10. **ECOLOGICAL INFERENCE** by Laura Irwin Langbein and Allan J. Lichtman
11. **MULTIDIMENSIONAL SCALING** by Joseph B. Kruskal and Myron Wish
12. **ANALYSIS OF COVARIANCE** by Albert R. Wildt and Olli Ahtola
13. **INTRODUCTION TO FACTOR ANALYSIS: What It Is and How to Do It** by Jae-On Kim and Charles W. Mueller
14. **FACTOR ANALYSIS: Statistical Methods and Practical Issues** by Jae-On Kim and Charles W. Mueller
15. **MULTIPLE INDICATORS: An Introduction** by John L. Sullivan and Stanley Feldman
16. **EXPLORATORY DATA ANALYSIS** by Frederick Hartwig with Brian E. Dearing
17. **RELIABILITY AND VALIDITY ASSESSMENT** by Edward G. Carmines and Richard A. Zeller
18. **ANALYZING PANEL DATA** by Gregory B. Markus
19. **DISCRIMINANT ANALYSIS** by William R. Klecka
20. **LOG-LINEAR MODELS** by David Knoke and Peter J. Burke
21. **INTERRUPTED TIME SERIES ANALYSIS** by David McDowall, Richard McCleary, Errol E. Meidinger, and Richard A. Hay, Jr.
22. **APPLIED REGRESSION: An Introduction** by Michael S. Lewis-Beck
23. **RESEARCH DESIGNS** by Paul E. Spector
24. **UNIDIMENSIONAL SCALING** by John P. McIver and Edward G. Carmines
25. **MAGNITUDE SCALING: Quantitative Measurement of Opinions** by Milton Lodge
26. **MULTIATTRIBUTE EVALUATION** by Ward Edwards and J. Robert Newman
27. **DYNAMIC MODELING: An Introduction** by R. Robert Huckfeldt, C. W. Kohfeld and Thomas W. Likens
28. **NETWORK ANALYSIS** by David Knoke and James H. Kuklinski
29. **INTERPRETING AND USING REGRESSION** by Christopher H. Achen
30. **TEST ITEM BIAS** by Steven J. Osterlind
31. **MOBILITY TABLES** by Michael Hout
32. **MEASURES OF ASSOCIATION** by Albert M. Liebetrau
33. **CONFIRMATORY FACTOR ANALYSIS: A Preface to LISREL** by J. Scott Long
34. **COVARIANCE STRUCTURE MODELS: An Introduction to LISREL** by J. Scott Long
35. **INTRODUCTION TO SURVEY SAMPLING** by Graham Kalton
36. **ACHIEVEMENT TESTING: Recent Advances** by Isaac I. Bejar
37. **NONRECURSIVE CAUSAL MODELS** by William D. Berry
38. **MATRIX ALGEBRA: An Introduction** by Krishnan Namboodiri
39. **INTRODUCTION TO APPLIED DEMOGRAPHY: Data Sources and Estimation Techniques** by Norfleet W. Rives, Jr. and William J. Serow
40. **MICROCOMPUTER METHODS FOR SOCIAL SCIENTISTS** by Philip A. Schrodt
41. **GAME THEORY: Concepts and Applications** by Frank C. Zagare